Common Poisonous Plants and Mushrooms

OF NORTH AMERICA

Common Poisonous Plants and Mushrooms

OF NORTH AMERICA

Nancy J. Turner, Ph.D.

and

Adam F. Szczawinski, Ph.D.

TIMBER PRESS
Portland, Oregon

This book cannot replace the advice and assistance
of qualified medical personnel.
In all cases of suspected poisoning by plants, or any other substance,
immediate qualified medical advice and assistance should be sought.

© 1991 by Timber Press, Inc.
All rights reserved.

Second printing 1992

ISBN 0-88192-179-3
Printed in Singapore

TIMBER PRESS, INC.
9999 S.W. Wilshire
Portland, Oregon 97225

Library of Congress Cataloging-in-Publication Data

Turner, Nancy J., 1947-
    Common poisonous plants and mushrooms of North America / Nancy J.
Turner and Adam F. Szczawinski.
        p.   cm.
    Includes bibliographical references.
    Includes index.
    ISBN 0-88192-179-3
    1. Poisonous plants--Identification. 2. Poisonous plants-
-Toxicology. 3. Mushrooms, Poisonous--Identification.
4. Mushrooms, Poisonous--Toxicology. 5. Poisonous plants--North
America--Identification. 6. Mushrooms, Poisonous--North America-
-Identification.   I. Szczawinski, Adam F. II. Title.
QK100.A1T87 1991
581.6'9'097--dc20   74282   90-37574
                                                        CIP

CONTENTS

IN CASE OF POISONING

For easy reference, write in the name and phone number of family physician and local Poison Control Center in the space provided.

Doctor: _____

Number: _____

Poison Control Center: _____

Number: _____

If a poisonous or unknown plant (or other poisonous substance) has been swallowed:

- **Identify** the plant or substance as closely as possible.

- Call your local **Poison Control Center** immediately, and seek professional **medical aid** and advice.

- **Do not panic; keep patient as comfortable as possible;** remember that there are very few really serious poisonings from plants, and even fewer fatalities.

- **Keep a sample** of the plant or substance, and if vomiting has occurred, save a sample of the vomit for further identification.

- **Watch breathing** closely, and if necessary, apply artificial respiration (CPR).

- **DO NOT induce vomiting** without medical help, or give anything to drink **IF** patient is extremely drowsy, unconscious, or convulsing, or if patient is under 12 months old.

- **DO Induce vomiting IF** patient is conscious, over 12 months old, and has swallowed a plant known to be poisonous, or is showing symptoms from swallowing an unknown plant. **Ipecac syrup** is the emetic of choice (see p. 10); follow dosage on bottle; usually works in 20–30 minutes. Lacking that, liquid household detergent (2–3 tablespoons in ½ to 1 cup of juice or water; or 30–45 ml in 125–250 ml juice or water) may be useful. Drinking a solution of table salt, a mixture of mustard and water, or "finger-down-the-throat" are not recommended for inducing vomiting (usually incomplete or unsuccessful). Save vomit for laboratory.

- **Gastric lavage** (removing stomach contents through a tube, adding fluids to help wash out the toxic materials) may be required if vomiting is unsuccessful or incomplete (see p. 11). This should be done **only** by a qualified physician.

- **Activated charcoal,** given in a slurry mixed with water **AFTER** syrup of ipecac has been given and patient has vomited, or after gastric lavage, can "tie up" toxic substances and prevent them from being absorbed by the body (see p. 11 for details). Repeat every 4–6 hours for 24 hours.

- **If patient is convulsing or clenching teeth,** place a piece of rolled cloth between the jaws to prevent damage to teeth and tongue.

- **For eye contamination** by plant irritants gently flush the eyes with a stream of slightly tepid water or saline solution for 5 minutes.

- **For skin contamination** by plant irritants, wash skin immediately with a large amount of running water; use soap if available; remove and thoroughly wash contaminated clothing.

- **If poisonous plant is known,** follow specific instructions for treatment.

**This book cannot replace the advice and assistance
of qualified medical personnel.**
**In all cases of suspected poisoning by plants, or any other substance,
immediate qualified medical advice and assistance should be sought.**

HOW TO PREVENT POISONING

- Never eat any part of an unknown plant or mushroom. If you are trying a "new food," eat only a small amount at first, and do not mix it with other "new" foods.

- Babies and young children must be carefully supervised. Keep poisonous house plants out of their reach. Remove poisonous berries from house and garden plants. Remove mushrooms from lawns and play areas. Store bulbs and seeds out of sight and out of reach.

- Teach children about poisonous plants just as they are taught about busy roads and hot stoves. Train them never to put any plants or parts of plants into their mouths. Even babies and toddlers can learn about "bad" plants.

- Learn to identify poisonous plants in the house, yard, and neighborhood. Eliminating all poisonous plants is not practical; knowing about them is.

- Teach children to recognize Poison Ivy, Poison Oak, and Poison Sumac or any other plants in the locality causing skin injuries, or dermatitis.

- Do not assume a plant is safe because birds or other wildlife eat it. What is poisonous to humans is not necessarily harmful to other animals.

- Do not assume, just because the fruits or roots of a plant are edible, that other parts of the plant can also be eaten. Many plants with edible parts also have poisonous parts.

- Do not count on heating or cooking a plant or mushroom to destroy any toxic substances.

- Avoid breathing or coming in contact with smoke from burning plants.

- Do not use unknown plants as skewers for meat or marshmallows, as decorations, or as playthings for children.

- Anyone using herbal medicines or teas, wild edible plants, or wild mushrooms, should be **positive** of the identifications of the plants they use and know how to prepare them properly. Each year many people are poisoned, some fatally, from mistaken identifications or misuse of herbal remedies.

- Do not use edible plants or mushrooms gathered from roadsides, or from areas where herbicides or insecticides may have been applied.

- Keep ipecac syrup on hand as part of any family first aid kit; keep enough for one dose for each family member. It is available at any pharmacy, and when used as directed is a safe, effective medicine for inducing vomiting (see p. 10). Powdered, activated charcoal is also good to keep on hand for use after vomiting to adsorb toxic substances in the digestive tract, but it is usually administered at a hospital (see p. 11).

**This book cannot replace the advice and assistance
of qualified medical personnel.
In all cases of suspected poisoning by plants, or any other substance,
immediate qualified medical advice and assistance should be sought.**

PREFACE

In April 1980, a five-year-old child was fatally poisoned in Victoria, British Columbia from eating Poison Hemlock (*Conium maculatum*) while at play with her sisters. Her babysitter was not even aware that she had eaten the plant. The little girl felt sick and would not eat. She lay down, and within an hour fell into a deep coma. It was only at this point that her sisters recalled that earlier she had eaten a plant. She was rushed to the hospital, but despite all efforts to save her life, she died six days later. In February 1987, under similar circumstances a six-year-old, also of Victoria, became paralyzed and stopped breathing after eating Poison Hemlock, but hospital staff were able to save her life by using a respirator and pumping out her stomach.

These two poisoning incidents are only two out of hundreds in which humans have been poisoned from eating plants or drinking teas made from plants which are harmful. Many involve young children at play, "pretending" to eat better-known, edible look-alikes.

Fig. 1. Poison Hemlock (*Conium maculatum*) leaves are sometimes mistaken for Parsley.

Fig. 2. Children are sometimes poisoned from the carrot-like taproots of Poison Hemlock.

In fact, statistics show that children aged five and under are most vulnerable to accidental poisoning. The pea-like pods of the Golden Chain tree (*Laburnum*), the onion-like bulbs of Daffodil (*Narcissus*), the carrot-like taproots of Poison Hemlock, and the attractive, colorful berries of Climbing Nightshade (*Solanum dulcamara*), Daphne (*Daphne mezereum*), and Lily-of-the-Valley (*Convallaria majalis*) have all severely poisoned children.

It is impossible to remove poisonous plants from our environment. The best way to prevent poisoning is for people to learn which plants are harmful and to teach their children not to play with plants and never to put them in their mouths. We hope that this book will be useful to parents, pet owners, outdoors people, gardeners and people who keep house plants, and to health care workers, helping them to identify the poisonous plants around them, and helping them to undertake, or at least understand the treatments for plant poisoning. In writing the book, we have tried to avoid the use of technical jargon, either botanical or medical, but at the same time to provide information that can be used in identification and treatment of plant poisonings.

For the more technical aspects of plant identification and treatment, and for descriptions of plants known to be poisonous to livestock but not to humans, other references are provided, many of which contain complete literature citations for various plant species and families and groups of plant toxins. If this book can prevent a single poisoning or result in a single successful treatment of a plant poisoning, we will feel it was worth our efforts.

In all cases of suspected poisoning by plants or any other substance, medical advice should be sought from the family physician or local Poison Control Center. This book is not intended to replace the advice of trained medical personnel, but to provide general information about potentially poisonous plants. The treatments listed for different types of plant poisoning are provided as guidelines only, in case of emergency when medical advice is not immediately available, or to give a better understanding of treatment for poisoning undertaken by physicians or hospital staff.

**This book cannot replace the advice and assistance
of qualified medical personnel.
In all cases of suspected poisoning by plants, or any other substance,
immediate qualified medical advice and assistance should be sought.**

ACKNOWLEDGEMENTS

We are particularly grateful to Dr. Walter H. Lewis, Professor, Department of Biology, Washington University, St. Louis, Missouri for reading the manuscript and for many helpful suggestions and contributions. Dr. George H. Constantine, Associate Dean and Head Advisor, College of Pharmacy, Oregon State University, Corvallis read the introductory section and gave valuable advice. Dr. Bettina Dudley and Dr. T. R. Dudley are also gratefully acknowledged for their editorial and botanical contributions.

The section on poisonous mushrooms was read by Dr. Robert Bandoni, Professor Emeritus, Department of Botany, University of British Columbia, Vancouver, Dr. Scott A. Redhead, Biosystematics Research Centre, Agriculture Canada, Ottawa, Ontario, and Dr. Al Funk and John Dennis of the Pacific Forest Science Research Centre, Victoria, B.C., all of whom offered many suggestions and helpful comments. Dr. Glen S. Jamieson and J.N.C. (Ian) Whyte, Pacific Biological Station, Nanaimo, B.C. reviewed the section on algae in Chapter III and provided important information on algal toxins.

The following people also contributed to this book in a variety of important ways: Dr. Eugene Anderson, University of California, Riverside; Dr. Denis R. Benjamin, Director of Laboratories, Children's Hospital and Medical Center, Seattle; Dr. Philip Chambers, M.D., of Nanaimo, B.C.; Dr. Adolf Ceska and Dr. Richard J. Hebda, Royal British Columbia Museum, Victoria, B.C.; Ken O'Connor, Victoria, B.C.; Dr. Amadeo Rea, San Diego Natural History Museum; Dr. R. E. Schultes, Director Emeritus, Harvard University Botanical Museum, Cambridge, Massachusetts; Krista Thie, White Salmon, Washington; and Dr. G. N. Towers, Department of Botany, University of British Columbia. These people are gratefully acknowledged.

Credits for the color illustrations are provided with the captions. Contributors include Mary W. Ferguson, Dr. Richard J. Hebda, Dr. Walter H. Lewis, Frank Fish, Ian G. Forbes, Dr. J. C. Raulston, Robert A. Ross, Dr. Al Funk, John Dennis, Kit Scates Barnhart, Dr. Eugene N. Anderson, Bill McLennan, Biosystematics Research Centre, Ottawa, Washington University, St. Louis, and the late Audrey Burnand (Royal British Columbia Museum collection). The remaining illustrations are by Robert D. and Nancy J. Turner.

We have enjoyed our partnership with Richard Abel and the staff of Timber Press in producing this book, and would like to thank them for their guidance and enthusiasm.

Finally, without the continuing help and support of our families, including Robert D. Turner, Jane Chapman, Isobel Turner, and Barbara Lund and Alan Szczawinski, the book would not have been possible.

HOW TO USE THIS BOOK

The Contents show the overall organization of the book. In Chapter II, poisonous mushrooms are listed in alphabetical order by scientific name, since their common names are variable and their scientific names often well known. For the succeeding chapters, within the broad categories indicated in the Contents, the plants are listed alphabetically by their prevalent common names, with alternate common names, scientific names, and family names also provided. In cases where a plant could potentially be included in more than one category, for example either as a wild plant or a garden plant, we have tried to place it in its most relevant context. If there is any doubt about where to find a reference to a particular plant, or if only the scientific name is known, please consult the Index. Technical terminology has been kept to a minimum. Some specialized botanical terms are defined in the Glossary (p. 281).

**This book cannot replace the advice and assistance
of qualified medical personnel.
In all cases of suspected poisoning by plants, or any other substance,
immediate qualified medical advice and assistance should be sought.**

CHAPTER I

AN INTRODUCTION TO POISONOUS PLANTS

RECOGNIZING POISONOUS PLANTS AND THEIR EFFECTS

What Is a Poisonous Plant?

Plants and parts of plants that contain potentially harmful substances in high enough concentrations to cause chemical injury if touched or swallowed are known as "poisonous." There are literally thousands of poisonous plants in the world. In this book are described over 150 of the most common and dangerous types to be found in the temperate regions of North America: in homes and buildings, in gardens and urban areas, and in the wild. Emphasis is given to plants known to have caused poisoning to humans, but many of the same types have also caused injury to animals, including household pets and grazing or browsing livestock.

It is important to remember that "natural" and "organic" do not necessarily mean "safe" and "wholesome." Some of the most virulent poisons known are derived from plants, which are, of course, both "natural" and "organic." Many medicinal plants, and even many food plants, contain chemicals that can be harmful. Our bodies are adept at handling and eliminating small amounts of many potentially harmful substances. In large concentrations, however, these same substances, when ingested, can affect our digestive, circulatory, or nervous systems, can cause irreparable damage to our liver or kidneys, can lower blood sugar, interfere with normal blood clotting, prevent cell division, or affect our immune systems. Other substances found in plants can cause skin reactions—pain, redness, blistering, swelling—or can harm the eyes, simply from contact.

Poisonous substances in plants are often considered to be by-products of essential plant functions. For many of these chemical compounds, scientists have not determined any obvious role in the functioning, or metabolism of the plants containing them. Some may simply be waste products. As well as being poisonous, many such compounds are bad tasting or have an unpleasant odor. These give protection to the plant against browsing animals and plant-eating insects.

Often, individual plants or local populations of plants vary in the relative concentrations of toxic compounds they contain. This natural variation has helped plant breeders over many centuries to select and develop plant varieties with low concentrations of poisonous substances. In fact, some of the vegetables we enjoy today were derived from ancestor plants too toxic to eat in any quantity. The potato is a good example. Even the edible tubers contain traces of the bitter alkaloid solanine, which is found in higher concentrations throughout the potato plant, rendering the green leaves, sprouts, and green, light-exposed tubers very poisonous. Tubers of the potato's wild ancestors are also poisonous. It was only through careful and continued selection of less toxic potato varieties that the native farmers of the South American highlands were able to develop such a delicious vegetable tuber. Johns and Kubo (1988) describe some of the means people have devised to render otherwise poisonous plants edible.

Can a plant be called poisonous if it has never poisoned anyone? The answer is yes. If it contains known toxic compounds in concentrations high enough to be dangerous if the plant is consumed, then it should be considered poisonous. Some plants can only be inferred to be toxic for humans, either because they are known to be poisonous to some animal species, or because chemical studies have revealed the presence of dangerous compounds. Controlled feeding studies with laboratory animals are sometimes used to determine potential toxicity, although many people feel this is an unacceptable and inhumane means of learning such information. A number of house plants previously not known to be poisonous, for example, were recently identified as apparently toxic by laboratory testing. At least one *Begonia* hybrid, *Gladiolus gandavensia* flowers, species of Dragon Tree (*Dracaena*), Prayer Plant (*Maranta*), *Peperomia*, Snake Plant (*Sansevieria*), *Selaginella*, and Goosefoot Plant (*Syngonium*), all received positive toxicity scores (Der Marderosian and Roia 1979). Because there is little other evidence against these plants, they are not described in detail in this book, but people should be wary of them nonetheless.

Herbicides and pesticides used on house and garden plants or in forests can render otherwise edible species dangerous, causing unpleasant reactions that may not be distinguishable from plant poisoning. Never harvest plants, berries, or mushrooms along roadsides or highway rights-of-way, because toxic heavy metals from vehicle exhaust, such as lead and cadmium, can accumulate in plant structures and mushrooms. People seeking aquatic or marine edible plants such as Cattail (*Typha latifolia*), Watercress (*Nasturtium officinale*), or seaweeds should ensure that there are no pollutants in the waters where the plants are harvested. Some molds and mildews growing on foods can also render them poisonous due to the presence of mycotoxins (see p. 65); moldy or rotting foods should be strictly avoided, and foods such as nuts and grains should be stored in cool, dry places to reduce the chances of them being infected with harmful fungi or bacteria.

Some plants, including many range species that might normally be edible for animals, have the capacity to absorb and concentrate naturally-occurring compounds such as nitrates, selenium salts, lead, and molybdenum. This sometimes leads to large-scale poisoning of livestock. Beets, Turnips, and Kale, for example, may accumulate excessive amounts of nitrates if they are over-fertilized. The nitrates may then be converted by bacteria in the digestive tracts of livestock to much more toxic nitrites. Since relatively large amounts of the plants must be consumed for toxic reactions to occur, and because many of the plant species involved are not normally eaten by people, human poisoning has seldom been a problem in these cases. The reader is referred to Kingsbury (1964) and Fuller and McClintock (1986) for information on this type of poisoning.

Prevalence of Poisoning by Plants

Poisonous plants are everywhere. They are found among virtually all groups of plants—algae, lichens, mushrooms, ferns, coniferous trees, and flowering plants. Many of our most prized and admired garden flowers, ornamental trees, and house plants are poisonous and some can be deadly.

Each year, throughout North America, newspapers and medical journals describe hundreds of instances of poisoning from ingestion of plants and plant parts. Babies and toddlers are often involved, especially in poisoning by ornamental house plants such as Dieffenbachia and Philodendron. Older children at play may eat poisonous plants, in imitation of real foods. Adult poisonings generally result from misidentification of "edible" wild plants, or misinformed use of herbal remedies.

Many instances of poisoning may not even be reported. As authors of books on edible wild plants and lecturers on this subject, as well as in interviewing elderly Native people, we hear many accounts of poisoning, some very serious, which are not necessarily officially documented. At least a dozen people on different occasions have told us they have tried to eat Skunk-Cabbage (*Lysichitum americanum*) leaves because they "looked good" or "must be like cabbage." The resulting intense and prolonged burning of the lips and mouth usually prevents ingestion of more than a bite, but one man was

Fig. 3. Poisonous plants such as Dieffenbachia and Philodendron are often used as ornamentals in shopping malls, offices, and lobbies.

hospitalized for several days, and experienced a swollen, raw throat "as if I had swallowed a whole cup of scalding coffee." Dieffenbachia, or Dumbcane, a relative of Skunk-Cabbage, is one of the most commonly ingested houseplants, and we have heard of at least three instances of adults, who should know better, inadvertantly chewing on the stems. Another incident, which could have been disastrous but fortunately was not, occurred when a man attending a banquet, in a fit of boredom over the after-dinner speech, reached out and plucked a Daffodil out of a vase in the center of the table, and ate it. He assumed it was edible—a foolish assumption indeed.

Some Native people, and more than one outdoor recreationist, have been seriously, sometimes fatally, poisoned from mistaking the highly toxic Water Hemlock (*Cicuta* spp.) for the similar-looking Water Parsnip (*Sium* spp.). Others have been poisoned from eating an overdose of the deadly Indian Hellebore (*Veratrum viride*) or Death Camas (*Zigadenus venenosus*), which are Indian medicinal remedies. We heard of a case where a hospitalized pregnant woman, anxious to proceed with her delivery and return home to her older children, secretly took a preparation of Blue Cohosh root (*Caulophyllum thalictroides*) to bring on labor. She had misjudged the age of her unborn baby, and it was delivered prematurely, at great risk to both mother and infant.

In the past, human poisonings occurred in epidemic proportions from such substances as Ergot (*Claviceps* spp.) and milk from cows feeding on White Snakeroot (*Eupatorium rugosum*). These types of poisonings have been virtually eliminated, but at the same time more and more people are seeking recreation in the out-of-doors, moving to suburban environments, and bringing new and exotic plants into their homes and gardens. The increasing popularity of herbal medicines and teas and of eating wild plants has resulted in a greater exposure of people to potentially harmful plant substances, and hence, despite more advanced knowledge and education on toxic plants, poisonings from plants continue to plague us. Fortunately, despite the thousands of cases of plant poisonings that occur in North America each decade, very few actually result in serious illness or death.

Plant Allergens and "Hay Fever"

Some types of plants and plant products cause allergic reactions in people and animals. This is when individuals have or develop unusual sensitivities to substances that may be harmless to others. Allergy-causing materials, known as allergens, include airborne spores of fungi and pollen grains from many different plant species, which produce in susceptible individuals a condition widely known as "hay fever." Although hay fever is not usually life-threatening, it does cause cold-like symptoms including stuffed-up nose, running eyes, and sneezing, leaving hay fever sufferers miserable during flowering season. Furthermore, if left untreated, hay fever can lead to asthma and other serious complications. Many, many plants, including grasses and numerous members of the aster, or composite family, such as Ragweed (*Ambrosia* spp.), cause widespread allergies. It is not within the scope of this book to describe the many hay fever plants, but details can be found in *Airborne and Allergenic Pollen of North America* (Lewis et al. 1983). If hay fever is suspected, a physician or allergy specialist should be consulted.

Other plants can cause even more serious allergies in some individuals. Some people, for example, are violently allergic to Peanuts, Buckwheat, or any number of other specific types of plants and plant products. These allergies can be life-threatening for some, because reactions to ingesting even minute quantities of an allergen can be severe, with vomiting, swelling of the throat, and restricted breathing. For people suffering from an allergy of this type, even certain common and normally edible plants are deadly poisonous. It is impossible here to deal with such plant allergens because they are so diverse and so specific to individuals.

Another common type of allergy is caused by skin contact with certain plants by individuals sensitive to them. The best known contact plant allergens are probably Poison Ivy and its relatives, Poison Oak and Poison Sumac. According to Hardin and Arena (1974), one out of every two people is allergic to these plants to some degree, and for some the reaction is severe enough to require hospitalization. These plants are described here in detail (Chapter III), and a number of others that commonly cause injury to the skin (dermatitis), through allergic reaction or chemical irritation are listed in Appendix 3. For a more thorough treatment of plants causing skin allergies and of other types of plants injurious to the skin, the reader is referred to Mitchell and Rook (1979).

Identifying Poisonous Plants and Mushrooms

The best time to identify plants or mushrooms that might be poisonous is before any poisoning has occurred. Correct identification can be critical, and people tempted to harvest wild plants or mushrooms for eating should take exceptional care to identify, or have an expert identify, any plant before it is consumed. Parents of young children, pre-school teachers, or anyone else caring for youngsters should also take special precautions to identify ornamental plants they keep indoors or have planted in yards. Never buy a plant for the house unless its identity is known, and its potential for harm understood. Florists and plant nursery workers may not volunteer information about potential toxicity of the plants they sell, but if asked, they will provide this information or at least the name of a plant so that the buyer can find out more about it. Kingsbury (1979) discusses the problems of labelling cultivated toxic plants and educating people about them.

Fig. 4. Poisonous plants are sometimes mistakenly eaten for edible species. Here, Death Camas (*Zigadenus venenosus*), with cream-colored flowers, is growing together with the edible Blue Camas (*Camassia quamash*), whose bulbs were an important food of Indian peoples of the Pacific Northwest.

There are many other places where information about plant identification and poisonous plants can be obtained, and many people who are willing and able to help. Libraries, nature centers, natural history museums, universities, medical centers, poison control centers, veterinary clinics, forest research facilities, horticultural centers, and botanical gardens are all places where people can go to find out more about poisonous plants. Many of these institutions employ plant experts (botanists) who are pleased to help people identify plant specimens. Many communities and institutions offer courses in plant or mushroom identification for the non-specialist, and these are an excellent way to learn the basics of plant identification—and to avoid accidental poisoning by toxic plants.

Poisoning of Animals by Plants

Although this book concentrates on human poisoning by plants and mushrooms, most of the species described are known to be toxic to animals as well.

Household pets, especially puppies and kittens, are as vulnerable to poisoning from plants as young children. In the fall of 1988 in British Columbia, for example, at least ten dogs and cats were severely poisoned, two of them fatally, from eating poisonous mushrooms in their owners' yards. Pet owners should watch their animals closely around dangerous plants, whether indoors or out. If poisoning is suspected prompt attention, just as with a child, can mean the difference between survival and death or lasting damage. As with human poisoning, samples of the plants or mushrooms eaten, or vomited materials, can aid significantly in the identification process, and hence, in the treatment given. Dogs and cats known to have eaten poisonous plants or showing symptoms of poisoning can be given syrup of ipecac to induce vomiting, just as in humans. Horses, cattle, sheep, and goats, however, do not ususally respond to the use of emetics.

Different types of animals have different levels of tolerance to poisonous plants. Cattle, sheep, goats, and other ruminants can often eat larger quantities of toxic plants without noticeable symptoms than single-stomached animals, such as horses, pigs, dogs, cats, and humans. Some animals have developed special enzymes than enable them to break down some toxic compounds into harmless ones. For example, some rabbits possess an enzyme that allows them to eat Belladonna (*Atropa belladonna*) in quantities that would be fatal to other animals. Squirrels have often been observed eating poisonous *Amanita* mushrooms, and deer are said to be able to feed on Yew and Rhododendron.

Because of their general unpalatability, many poisonous plants are avoided by browsing or grazing animals except in times of extreme hunger and food shortage, which, for range animals, often occur in the winter and early spring months. At such times an animal may develop a taste for a poisonous plant not normally eaten, and become addicted to it, still seeking it when other feed is available. Hence, in animals once poisoned there are often recurring episodes, and this can be a serious problem for farmers and ranchers.

A major problem in animal poisoning is that it often takes place away from human observation, and the only indication that poisoning has occurred is when symptoms appear. Since most poisonous plant manuals classify by plant species rather than by exhibited symptoms, and since many different toxic plants produce similar symptoms, it is often difficult to pinpoint which plant or plants have caused the problem and what the best treatment would be.

Some types of animal poisoning are chronic and difficult to detect. They are manifested in high rates of miscarriage, stillbirths, and infant mortality, lowered milk or wool production, slow growth, and general listlessness. All of these result in severe economic losses to farmers and ranchers as well as untold suffering to the animals.

The best means of preventing acute or chronic poisoning in animals, as well as obtaining the healthiest and most productive livestock, is to maintain high quality pasturage and grazing lands, provide supplemental feeding whenever required, ensure adequate intake of essential minerals such as calcium and magnesium, and offer a good supply of fresh, clean water. Overgrazed lands lead to underfed livestock far more likely to consume any poisonous plants available to them. For further information on animal poisoning by plants, the reader is referred to Fowler (1980), Hulbert and Oehme (1984), and Keeler et al. (1978).

Poisonous Plants Can Be Useful

Just because a plant is considered poisonous or harmful in some way does not mean that it cannot be useful or valuable. Many poisonous plants are beautiful and decorative and can provide "food for the soul," if not for the body. Aside from their ornamental value, though, many provide important compounds widely used in traditional and modern medicine. The very actions that render them poisonous—their effects on the heart and circulatory system, the nervous system, the blood, or on cell division for example—make them useful in treating certain conditions. Pharmacognosy is the science dealing with the knowledge of natural drugs and their biological, chemical, and economic features. By definition, this science also includes the study of the poisonous properties of medicinal plants.

Foxglove, or Digitalis (*Digitalis purpurea*) is one of the best known and most widely used poisonous medicinal plants. It was known to be poisonous but was used for centuries by illiterate farmers and housewives in England and Europe for treating "dropsy," a condition of massive fluid retention in the body, known more recently to be brought about by poor heart function. The link between Foxglove, heart activity, and dropsy was first proposed in the late 1700s by a prominent British physician, William Withering. We now know that Foxglove contains a mixture of cardiac glycosides, including digitoxin, gitoxin, and gitaloxin, which have a strong effect on the heart muscles. When used in correct dosages they change the rhythm of the heart beat, lengthening the time between heart contractions and thus allowing the ventricle to be emptied more completely. They also improve general circulation, relieve fluid buildup, or edema, and help kidney secretion. The dosage must be carefully controlled, however, because the effective therapeutic dose may be as high as 70 per cent of the toxic dose. Today, millions of people take digitalis to help regulate and strengthen their heartbeat—proof that poisonous plants can be life-savers as well as life-takers.

There are many other examples of useful, yet toxic plants. Opium Poppy (*Papaver somniferum*) yields, among other compounds, the potent but habit-forming pain killer, morphine, as well as the milder, usually non-habit-forming pain relief medicine, codeine. Both are alkaloids, isolated from opium. A third opium alkaloid, papaverine, is used mainly in the treatment of internal spasms, particularly of the intestines.

Indian Hellebore (*Veratrum viride*), which has been used medicinally by North American Indians for generations, yields a number of important alkaloids, particularly the ester alkaloids germidine and germitrine, which are used in the treatment of high blood pressure.

Mayapple (*Podophyllum peltatum*) contains lignans having anti-cancer and anti-viral properties, and today provides a drug of choice for treating human venereal warts. Autumn Crocus (*Colchicum autumnale*) has been investigated for use in cancer chemotherapy because one of its alkaloids, colchicine, interferes with cell division and hence with the proliferation of rapidly growing cancer cells.

Many other toxic plants have been used in folk medicine in various parts of the world, and some are still marketed in health food and herbal healing stores. Most are safe if used correctly, but a very few products that have been bought and used as directed have resulted in serious illness or death for the user. Many herbs sold as teas or for medicinal use are labelled only by common names on their packages, and this can be misleading, since the same common name may be used for two or more distinct types of plants. Sometimes, too, people have a violent allergy to herbal medicines that may be safe for others. Since laws to guide the use of herbal medicines are not well established, the onus is on the user to ensure that the herbal medicines taken are safe. Some commonly used herbal medicines that are considered by experts to be generally unsafe or questionable are listed in Appendix 6. Further information on the role of toxic and other plants in medicine can be learned from Claus et al. (1970), Lewis and Elvin-Lewis (1977), Brinker (1983), and Tyler (1987).

TREATMENT

**This book cannot replace the advice and assistance
of qualified medical personnel.
In all cases of suspected poisoning by plants, or any other substance,
immediate qualified medical advice and assistance should be sought.**

Ipecac Syrup: A Safe, Effective Emetic

Ipecac syrup, or syrup of ipecac, is obtained from a plant in the coffee family, *Cephaelis ipecacuanha*. It contains several alkaloids that cause vomiting. Although it could, for this reason, be termed poisonous, it can be extremely useful in expelling a poisonous substance after it is swallowed and before it is assimilated by the body. Given orally, in a usual dose of 2 tablespoons (30 ml) for adults and 1 tablespoon (15 ml) for children 1 to 12 years old, it is a standard item in poison control kits, and should be kept on hand in any household, and particularly where there are young children. If there is more than one child in the family, or if there are young friends and neighbours, it is best to keep several doses available at all times, since children playing together may eat the same toxic plant at the same time.

Vomiting is usually slightly delayed, occurring from 15 to 30 minutes after the syrup is given. When used as recommended, ipecac is safe and fairly reliable. Be sure to keep the patient's air passage clear during vomiting; young children should be held in the "spanking position." Induced vomiting should be followed up with a drink of *clear* fluids (not milk): ½ to 1 cup (125 to 250 ml) for children up to 6 years of age, and 1 to 2 cups (250–500 ml) for patients 7 years of age or older. If vomiting does not occur after 20 to 25 minutes, the dose of ipecac syrup may be repeated *once*. If no vomiting occurs after two doses, gastric lavage should be performed by a physician.

Ipecac syrup should not be given in cases of ingestion of corrosive acids or alkalis such as lye, or if the patient is extremely drowsy or unconscious, or has lost the gag reflex. Infants under 6 months of age should not be given an emetic, and babies from 6–12 months only under supervision of an accredited health care worker. It may be ineffective for patients who are chronically under anti-emetic medication. Cats, dogs, and pigs can be given ipecac syrup safely, but it should not be administered to horses, cows, or other livestock.

Gastric Lavage, or Washing the Stomach

This procedure should be done *only* by a physician or qualified medic, and is usually carried out in cases where inducing vomiting is not possible or has been unsuccessful or incomplete. If an infant is poisoned, or if a patient is extremely drowsy, unconscious, has lost the gag reflex, or is convulsing, gastric lavage is usually indicated. It involves the passage of a tube into the stomach, removal of the stomach contents through gravity or suction, and careful replacement of the contents with water or normal saline solution. The procedure is usually repeated until the washings are clear and free of toxic materials, and is generally followed by a dose of activated charcoal and a saline cathartic. Unless the patient is in a coma, sedation is usually required before gastric lavage is carried out, because it is an uncomfortable and somewhat frightening procedure. It has saved many lives, however, and is a standard procedure in most cases of serious poisoning.

Activated Charcoal: A Natural Cleanser

Activated charcoal is a powerful, non-toxic adsorbing agent effective for many poisons. It is not effective against cyanide, caustic or corrosive substances, iron or lithium salts, lead, or ethyl or methyl alcohol. It should be administered *AFTER* ipecac syrup has been given and the patient has vomited, or after gastric lavage. It binds with toxic substances and prevents them from being absorbed by the body. It can even "catch up" with poisons along the digestive tract. It is usually given orally in a slurry with water. The usual dose is 2 to 3.5 ounces (50–100 g) in a glass of water for an adult, and 1/3 to 1 ounce (10–25 g) for a child, or about 1 tablespoon (15 ml) for every 17 pounds (8 kg) of body weight. The dose should be repeated every 4 to 6 hours for 24 hours. Charcoal tablets or burnt toast are not effective substitutes for activated charcoal. Activated charcoal is usually administered at a hospital and is not a home remedy, but it can be purchased and kept as part of a family's first aid supplies.

Cathartics

Cathartics—substances that speed up elimination of materials through the digestive tract—can be very useful in treatment of poisoning, but should be given only under medical supervision. They are usually given at the same time as activated charcoal: after the patient has vomited, or after gastric lavage has been performed. By hastening the passage of toxic materials through the digestive tract, they help to decrease the body's absorption of harmful compounds. They also help to disperse the activated charcoal more evenly and prevent it from becoming cemented into hard lumps. Cathartics include such substances as magnesium sulfate (Epsom salts), sodium sulfate, sodium phosphate-biphosphate complex (Fleet Enema®), and sorbitol. These can be given orally or through a tube. The first three are salts which are dissolved in water, juice, or sweetened fluid. The last can be given straight or mixed with activated charcoal. A table of dosages is given by the Canadian Pharmaceutical Association (1984). Cathartics should *not* be used for poisoning by corrosive substances or if a patient's electrolyte balance is disturbed. If the patient is suffering from kidney failure, magnesium-containing cathartics should not be used. Castor oil or mineral oil are not recommended because they may increase absorption of fat-soluble poisons.

Cyanide Poisoning and Its Treatment

Cyanides are produced by many plants from the breakdown of cyanogenic glycosides. Since cyanide poisoning can be rapid and catastrophic, a summary of its treatment is included here. Cyanide-producing compounds are found in the leaves, bark, and seeds of many edible fruits in the rose family (Cherries, Peaches, Apples, Pears, Apricots, Plums, Almonds), as well as in Elderberry plants (*Sambucus* spp.), Vetches (*Vicia* spp.), Sorghums (*Sorghum* spp.), some strains of Lima Bean (*Phaseolus lunatus*), Flax (*Linum usitatissimum*), Yew (*Taxus* spp.), Bracken Fern (*Pteridium aquilinum*), and Arrow-grass (*Triglochin* spp.).

The digestive system is capable of breaking down small quantities of cyanides, but in larger doses they cause anxiety, confusion, dizziness, headache, and vomiting, with an odor of bitter almonds on the breath or vomit. Difficult breathing, an initial increase in blood pressure and low heart rate, followed by high blood pressure and rapid heart rate, and possible kidney failure may also occur. Coma, convulsions, and death from respiratory arrest may happen suddenly and rapidly.

Treatment: If no symptoms are seen in the patient, observe closely for two hours. The initial treatment for serious cases must be rapid.

Maintain breathing, and give 100% oxygen initially. Use Lilly Cyanide Antidote Kit if available, making sure the expiration date is not exceeded. Administer amyl nitrite by inhalation for 30 seconds every minute while an intravenous injection of sodium nitrite is being prepared. Then, for an adult, inject 300 mg sodium nitrite at 2.5–5.0 ml per minute; for a child, 10 mg/kg should be the initial dose. Through the same needle, inject sodium thiosulfate over about 10 minutes. The adult dose is 12.5 g, and for a child, 1.65 ml/kg of a 25% solution. Check blood hemoglobin and methemoglobin levels. Further administration of sodium nitrite and sodium thiosulfate, especially in children, must depend on hemoglobin level. The Canadian Pharmaceutical Association (1984: 127) presents tables for sodium nitrite and sodium thiosulfate doses according to hemoglobin concentrations.

After the initial treatment, gastric lavage may be performed; it is seldom practical initially because of the rapid onset of life-threatening symptoms. Respiration, heart beat, blood (arterial gases, pH, lactic acid), and kidney function should be monitored. Convulsions may be treated with i.v. diazepam, and lactic acidosis with i.v. sodium bicarbonate. One hundred percent oxygen should be given to maintain high oxygen tension.

General Treatment for Digestive Tract Irritation Caused by Plants

Irritation of the mouth, throat, stomach, and intestines due to the ingestion of toxic substances is the most common type of internal poisoning by plants. Although the actual toxins and their means of action vary considerably, the general treatment for such irritation is relatively uniform. Suggested treatment procedures, as outlined by the Canadian Pharmaceutical Association's *Poison Management Manual* (1984), are provided here.

For mouth and throat irritation, caused by chewing on plants such as *Dieffenbachia* and other plants in the arum family (Araceae) containing calcium oxalate crystals that burn and irritate the mouth and throat, give milk or ice cream, or allow the patient to suck on a popsicle or ice chips. Observe for swelling of the mouth and throat which may block air passage. Antihistamines may relieve the swelling.

After ingestion of stomach irritants, including Wisteria (*Wisteria* spp.) and Daffodils (*Narcissus* spp.), vomiting usually occurs spontaneously. Perform gastric lavage if a potentially toxic amount was swallowed, and follow with activated charcoal and a saline cathartic. Antacids may provide relief. Maintain fluid and electrolyte balance, especially in cases of severe vomiting or diarrhea. Treat for shock, if present. Consult information on specific plant.

Ingestion of intestinal irritants usually results in vomiting, abdominal pain, and diarrhea within an hour. Other systemic effects occur only if large amounts have been ingested. In this group are saponin plants, such as Horse Chestnut and Buckeyes (*Aesculus* spp.), resin-containing plants such as Daphne (*Daphne* spp.), Iris (*Iris* spp.), and Pokeweed (*Phytolacca americana*), taxine-containing Yews (*Taxus* spp.), and miscellaneous plants including Oaks (*Quercus* spp.), Holly (*Ilex aquifolium*), Mistletoe (*Phoradendron* spp.), Poinsettia (*Euphorbia* spp.), Privet (*Ligustrum vulgare*), and Waxberry (*Symphoricarpos albus*). For these plants, induce vomiting or perform gastric lavage, and follow with activated charcoal and a saline cathartic. Maintain fluid and electrolyte balance. Treat for shock if necessary. Consult information on the specific plant.

Some plants cause intense digestive reactions which may be delayed for one hour and up to two days following ingestion. These include such violent toxins as colchicine in Autumn Crocus (*Colchicum autumnale*) and Gloriosa Lily (*Gloriosa superba*), oxalates and other toxins in Rhubarb (*Rheum rhabarbarum*), solanine in Nightshades, Potato sprouts and green tubers, and Jerusalem Cherry (*Solanum* spp.), and toxalbumins in Castor Bean (*Ricinus communis*), Rosary Pea (*Abrus precatorius*), and Black Locust (*Robinia pseudoacacia*). For these, consult recommended treatments for specific plants.

Treatment for Skin and Eye Irritants

Plants which irritate the skin and eyes are listed in Appendix 3, and some of the more serious dermatitis-causing plants (Poison Ivy, Poison Oak, and Poison Sumac) are treated in the text. There are various types of skin irritation, including mechanical injury caused by spines, thorns or barbs, chemical irritation, allergic skin reactions, and photodermatitis, which occurs when exposure to the plant is followed by exposure to sunlight. Irritation, itching, burning, blistering, and/or excessive pigmentation may result from exposure to these plants, and sometimes there is danger from secondary infection if the skin is broken.

Treatment for eye irritation should involve flushing the eyes with a gentle stream of tepid water; if the irritation persists an ophthamologist should be consulted.

For mechanical injury to the skin, remove thorns, spines, or barbs, and treat the wound with a disinfectant. Small cactus spines can sometimes be removed by dripping candle wax over the affected area, allowing the wax to harden by immersing the area in cold water, and then peeling off the hardened wax.

For other types of dermatitis, wash the skin thoroughly with soap and water. Remove any contaminated clothing and wash prior to reuse. For Poison Ivy or its relatives, application of calamine lotion, Burow's solution (containing aluminum acetate), or an antiperspirant containing aluminum salts may provide relief. Severe itching may be treated with diphenhydramine given orally or by muscular injection, or with topical or systemic corticosteroids. Close observation for bacterial or fungal infections should be maintained. Lampe and Fagerström (1968) and the Canadian Pharmaceutical Association (1984) can be consulted for further information on treatment of skin and eye irritations caused by plants.

TYPES OF PLANT POISONS

Classification of Plant Poisons

There are many types of potentially harmful substances to be found in plants. Biochemists classify them according to their chemical structures, as well as their effects on people and animals. Many poisonous plants contain more than one kind of toxic substance, and sometimes the major classes of toxic compounds overlap by virtue of sharing certain chemical structures. In spite of all that is known about the chemistry of toxic plants, there is still much more to learn. As already indicated, many plant toxins can be put to use by people as medicines, and undoubtedly many more medicinal uses are yet to be discovered.

Alkaloids and glycosides comprise two major classes of plant toxins, both highly complex and diverse. They are difficult to characterize in more than a general way, and in each case several subclasses are recognized based on chemical structure and physiological action. Some other poisonous principles found in plants are oxalates, tannins, phenols, alcohols, aldehydes, volatile oils, protein substances, and toxalbumins. Following is a brief overview of these classes. For further, more detailed information, the reader is referred to Fuller and McClintock (1986), Cooper and Johnson (1984), and Claus et al. (1970).

Alkaloids

Alkaloids are a major group of organic compounds. Over 4000 alkaloid compounds have been identified, and they occur in about 15 to 20 % of vascular plants and at least 40% of plant families. They are found in roots, seeds, leaves, bark, and stems. Many plant species contain several different alkaloids similar in structure. Alkaloids are mostly basic in nature, as implied by their name, meaning "alkali-like." Derived from amino acids, they have a complex molecular structure, with at least one nitrogen (N) atom included within a heterocyclic ring. By definition they have a specific pharmacological activity, especially on the nervous system of animals. This makes them both potentially toxic and potentially medicinal. Most are bitter-tasting.

Many alkaloids bear names based on the scientific name of their plant sources, with the ending -in, or -ine (e.g., solanine, from Nightshade, Solanum, and lupinine, from Lupine, Lupinus). Some are named from their physiological actions, such as morphine ("sleep inducing") from Opium Poppy, and emetine ("emetic acting") from the well known emetic Ipecac (Cephaelis ipecacuanha), described previously as the source of ipecac syrup.

Plant families with many alkaloid-containing species include the amaryllis family (Amaryllidaceae), aster or composite family (Asteraceae or Compositae), spurge family (Euphorbiaceae), bean or legume family (Fabaceae or Leguminosae), lily family (Liliaceae), poppy family (Papaveraceae), buttercup family (Ranunculaceae), madder family (Rubiaceae), and nightshade family (Solanaceae). Two very poisonous groups of alkaloids, however, are from Yew (Taxus spp.) in the yew family (Taxaceae) and Poison Hemlock (Conium spp.) in the celery or umbel family (Apiaceae or Umbelliferae); these families are otherwise poorly represented by alkaloids.

Examples of some major subclasses of alkaloids and representative toxic plants containing them are given in Table 1.

Once ingested, alkaloids may be chemically altered by enzyme reactions in the liver. Sometimes they are rendered harmless; in other cases they become even more deadly. This is apparently the case with the pyrrolizidine alkaloids found in Ragwort (Senecio spp.). It has been suggested that the Ragwort alkaloids themselves are not harmful to the liver, but compounds produced from them are. They cause irreversible liver damage in humans and animals for which there is no satisfactory treatment. People drinking herbal teas from Ragwort species, or even consuming milk or honey contaminated with Ragwort, are at risk even if the quantities consumed are minor.

Structural similarities have been shown between various alkaloids and nerve transmitter substances produced by the human body, including acetylcholine, norepinephrine, dopamine, and serotonin. It is thought that the toxicity of many alkaloids is due to their mimicking or blocking the action of such substances. Symptoms common to acute poisoning by these alkaloids include excess salivation, dilation or constriction of the pupils, vomiting, abdominal pain, diarrhea, lack of coordination, convulsions, and coma. Treatment is with drugs that counteract the central nervous system effects of alkaloids.

Table 1. Some major subclasses of alkaloids and representative plants containing them.

Alkaloid Subclass	Alkaloid Name	Toxic Plant Containing Alkaloid
pyrrolizidine	senecionine and others	Ragwort, Groundsel (*Senecio* spp.), Heliotrope (*Heliotropium* spp.)
piperidine	coniine	Poison Hemlock (*Conium maculatum*),
	lobeline	Indian Tobacco (*Lobelia inflata*)
tropane	atropine	Belladonna (*Atropa belladonna*)
	hyoscyamine	Henbane (*Hyoscyamus niger*)
	scopolamine	Jimsonweed (*Datura stramonium*)
quinolizidine	lupinine	Lupine (*Lupinus* spp.)
	cytisine	Laburnum (*Laburnum anagyroides*), Broom (*Cytisus scoparius*)
pyridine-piperidine	nicotine	Tobacco (*Nicotiana* spp.)
	anabasine	Tree Tobacco (*Nicotiana glauca*)
indole	psilocybin	Magic Mushroom (*Psilocybe* spp.)
	ergot alkaloids	Ergot (*Claviceps* spp.), Morning Glory (*Ipomoea tricolor*)
	gelsemine	Carolina Jessamine (*Gelsemium sempervirens*)
isoquinoline	berberine	Barberry (*Berberis* spp.)*
	emetine	Ipecac (*Cephaelis ipecacuanha*)
	protopine series	Opium Poppy (*Papaver somniferum*) and other Poppy species, Celandine (*Chelidonium majus*), Bloodroot (*Sanguinaria canadensis*)
steroidal	solanine	Nightshade, Potato (*Solanum* spp.)
	buxine complex	Box (*Buxus sempervirens*)
	germidine and others	Indian Hellebore (*Veratrum viride*)
	zygadenine	Death Camas (*Zigadenus* spp.)
diterpenoid	aconitine	Monkshood (*Aconitum* spp.)
	delphinine	Larkspur (*Delphinium* spp.)
purine	caffeine	Coffee (*Coffea arabica*)*, Tea (*Camellia sinensis*)*, Cocoa (*Theobroma cacao*)*, Holly (*Ilex* spp.)

*Apparently not highly toxic in moderation; not included, except in Appendix 1.

Glycosides

Glycosides are even more widely distributed in plants than are alkaloids. Many are non-toxic, but a significant number do yield poisonous compounds. The term glycoside is applied because all compounds in this class consist of one or more sugar molecules, combined with a non-sugar component, called an aglycone. When ingested, glycosides are readily broken down by enzymes or acids into sugar and aglycone units. The poisonous qualities of glycosides are determined by their aglycones, and the properties of the latter are often used to classify glycoside compounds.

For example, cyanogenic (or cyanogenetic) glycosides are compounds which release hydrocyanic acid as a by-product of their breakdown, which may occur when the plants are bruised, wilted, or ingested. Trace amounts of cyanogenic glycosides are found in many types of plants. Altogether, some 800 species in 80 different families contain them. However, relatively high concentrations leading to cyanide poisoning occur in fewer species, mainly in the rose family (Rosaceae) and bean or legume family (Fabaceae, or Leguminosae). Cyanide-producing plants possess an enzyme system which, when it comes in contact with the cyanogenic glycoside, breaks down the glycoside into sugar, cyanide, and an aldehyde or ketone. Cyanides inhibit oxygen uptake in the body's cells, and thus poisoning may occur very suddenly. (For symptoms and treatment of cyanide poisoning, see p. 12).

Another important group of glycosides is the goitrogenic glycosides, mustard oil glycosides, or glucosinolates. They are responsible for the hot, pungent flavor of radishes, cresses, cabbages, and other members of the mustard family (Brassicaceae, or Cruciferae). They may be present in all parts of a plant but the highest concentration is usually in the seeds. Goitrogenic glycosides break down with an associated enzyme into a glucose sugar, a sulfate fraction, and irritant mustard oils, or goitrogens, that interfere with the uptake of iodine by the thyroid gland and ultimately limit thyroxine production. In small quantities they are harmless to people eating a balanced diet, but if eaten in excess over long periods of time they could lead to lowered thyroid activity. Mustard oils should not be confused with "mustard gas," a name for the greenish-colored chlorine gas which has been used in wars as a chemical poison.

Cardiac, or cardioactive glycosides, are a group characterized for their direct action on the heart. Over 400 have been isolated, the best known being the digitalis glycosides present in Foxglove (*Digitalis purpurea*). Besides their effect on the heart muscle, cardiac glycosides can produce severe digestive upset with nausea, vomiting, abdominal pain, diarrhea, blurred and disturbed color vision, and other symptoms relating to decreased heart function.

Saponin glycosides have as their aglycone component a compound termed a sapogenin. They are water-soluble, bitter-tasting, and soap-like even at low concentrations. They are seldom harmful in small amounts, but if consumed in large doses their effects can be serious. Symptoms may include irritation of the mucous membranes, increased salivation, nausea, vomiting, and diarrhea, and in severe cases, dizziness, headache, chills, heart disturbances, and eventually convulsions and coma. Saponins alter the permeability of cell membranes and break down red blood cells; hence a saponin injected into the bloodstream can be fatal. They are highly toxic to cold-blooded animals and have been used around the world as fish poisons.

Examples of these and other major classes of toxic glycosides are given in Table 2.

Other Classes of Poisonous Plant Compounds

Plant toxins other than the alkaloid and glycoside classes form a diverse chemical mixture of many different types of compounds, some defined by their chemical composition and some by their effects on humans and other animals.

Oxalic Acid and Oxalates: Oxalic acid and its salts, called oxalates, occur in nearly all organisms, but in certain plant families such as the goosefoot family (Chenopodiaceae) and the buckwheat family (Polygonaceae) relatively large and potentially toxic amounts occur in a number of species. The sour taste of Rhubarb stalks and Sorrels is from the acid itself. Other plants contain high concentrations of the salts, including sodium oxalate and potassium oxalate, both water soluble, and calcium oxalate, which is not. When large quantities of plants containing oxalic acid or soluble oxalates are eaten, the oxalate component (oxalate ions) may combine with free calcium in the digestive tract to form insoluble calcium oxalate. By "tying up" the free calcium, this can lead to a calcium deficiency, especially if the diet is already poor in calcium. Additionally, insoluble calcium oxalate crystals may be deposited in the kidneys and other organs, causing mechanical damage.

Oxalate content varies considerably with the plant's age, and with seasonal, climatic, and soil conditions. Under normal circumstances, oxalates in moderate amounts can be broken down by bacteria in an animal's digestive tract. Usually oxalic acid salts must comprise 10 per cent or more of a plant's dry weight to be seriously toxic. Several range plant species in the goosefoot family are particularly notorious for their accumulation of large amounts of oxalates. These include Halogeton (*Halogeton glomeratus*), Greasewood (*Sarcobatus vermicu-*

Table 2. Some major classes of toxic glycosides and examples of plants containing them.

Glycoside Subclass	Glycoside Name	Toxic Plant Containing Glycoside
cyanogenic	amygdalin prunasin	Leaves, bark, and seed kernels of Apricot, Cherry, Plum (*Prunus* spp.)
	sambunigrin	Elderberry (*Sambucus* spp.)
	vicia group	Vetches (*Vicia* spp.)
goitrogenic	glycosinolates	Black Mustard (*Brassica nigra*) and other *Brassica* spp., Horseradish (*Armoracia rusticana*)
cardiac or cardioactive	digitalis group	Foxglove (*Digitalis purpurea*)
	cymarin	Indian-Hemp (*Apocynum* spp.)
	hellebrin	Hellebore (*Helleborus niger*)
	convallatoxin	Lily-of-the-Valley (*Convallaria majalis*), Star-of-Bethlehem (*Ornithogalum umbellatum*)
	evomonoside	Burning Bush (*Euonymus* spp.)
	oleandrin	Oleander (*Nerium oleander*)
saponin	githagenin	Corn Cockle (*Agrostemma githago*)
	aesculin	Horse Chestnut and Buckeye (*Aesculus* spp.)
coumarin-derivatives	umbelliferone	several genera of the celery family (Apiaceae, or Umbelliferae); Daphne (*Daphne* spp.)
	aesculin (see above)	Horse Chestnut and Buckeye (*Aesculus* spp.)
	daphnin	Daphne (*Daphne* spp.)
	furanocoumarins (including psoralens)	members of celery family (Apiaceae), aster family (Asteraceae, or Compositae), bean family (Fabaceae, or Leguminosae) (see Appendix 3 for examples)
protoanemonin	ranunculin	Buttercups (*Ranunculus* spp.), Anemones (*Anemone* spp., *Pulsatilla* spp.), and other buttercup family members
anthraquinone	barbaloin	Aloes (*Aloe* spp.)
	cascarin	Cascara (*Rhamnus purshiana*)*

*Used medicinally as a laxative; not generally considered toxic when used in moderation.

latus), and Russian Thistle (*Salsola* spp.). Of plants eaten by people, Rhubarb stalks (*Rheum rhabarbarum*) (the leaves contain other severe toxins), Beets (*Beta vulgaris*), Sorrels (*Rumex* spp., *Oxalis* spp.), and Purslane (*Portulaca oleracea*), should all be used only moderately and not in large amounts, due to their potential for accumulating oxalic acid and oxalate salts.

Additionally, certain plants, mostly in the arum family (Araceae), contain calcium oxalate in bundles of minute but sharp, needle-like crystals that cause intense burning, swelling, and discomfort to the lips, mouth, and throat if ingested, by puncturing the sensitive mouth tissues. This physical damage and irritation can become even more serious if the breathing passage is constricted by swelling. Many of our common house plants—Philodendrons, Dieffenbachias, Caladium, and Calla Lily, for example—are in the arum family and contain these irritating crystals. Several are discussed in detail in Chapter V.

Tannins and Other Phenols: Phenols are acidic, and form salts with alkaline compounds. Probably the most notorious types of phenols are those of the various species of *Toxicodendron* (also called *Rhus*) known as Poison Ivy, Poison Oak, and Poison Sumac, which cause serious allergic skin reactions in many people. Tannins are complex, astringent phenols, the active part being gallic acid. They are used commercially in tanning leather. They bind up proteins, including enzymes, and thus can quickly stop all cell functions. They occur in many tree barks and other plant structures, but are seldom a problem for humans. They are, however, present in the acorns and leaves of Oaks (*Quercus* spp.), making them extremely bitter and potentially toxic. Even edible types of acorns, before they can be safely consumed, usually must have their tannins removed through leaching.

Alcohols: These carbohydrate derivatives are best known to people in the form of ethyl alcohol, the active ingredient of wine, beer, and other alcoholic beverages that is produced by fermentation of carbohydrates. We do not normally consider this type of alcohol to be poisonous, but the first definition of "intoxication," the term normally applied for "being drunk," is ". . . essentially poisoning." Other types of alcohol can be even more deadly. Although toxic alcohols are rare in plants, Water Hemlock (*Cicuta maculata* and related spp.), considered probably the most violently poisonous plant of the North Temperate zone, contains a highly unsaturated higher alcohol, cicutoxin, as its main toxic principle. Another toxic alcohol compound is tremetol, which occurs in White Snakeroot (*Eupatorium rugosum*) and some of its relatives.

Aldehydes and Ketones: Aldehydes and ketones are carbohydrate derivatives, and are not commonly associated with plant poisoning. When a person has eaten Inky Cap mushrooms (*Coprinus atramentarius*), followed by an alcoholic beverage, coprine in the mushroom causes the body's metabolism of the alcohol to be slowed down. This results in the accumulation of a toxic aldehyde, acetaldehyde, and in the symptoms of coprine poisoning as described under this mushroom (p. 45).

Ketones are rarely toxic. One that is reputed to be harmful is found in European Pennyroyal (*Mentha pulegium*), making this mint potentially unsafe for use as a medicinal herb or tea.

Proteins, Peptides, and Amino Acids: Proteins are essential constituents of all living cells. They are extremely complex organic substances composed of infinite combinations of different types of amino acids linked together in various ways. Peptides are less complex combinations of two or more amino acids, and are usually produced through partial chemical breakdown of proteins. Longer peptide chains are known as polypeptides. Many different types of proteins, peptides, and amino acids occur in plants, and some are toxic.

Over 260 different amino acids are known, but only 20 occur in proteins. The others occur in a free state in living cells and are derived from chemical alteration of the protein amino acids. Certain amino acids called lathyrogens are found mainly in the seeds of Sweet Pea and its relatives (*Lathyrus* spp.) and some Vetches (*Vicia* spp.), and are responsible for a chronic disease called lathyrism (see p. 218).

Toxic polypeptides are relatively rare, but complex cyclic polypeptides are responsible for the most deadly types of mushroom poisoning. They are found in Death Cap (*Amanita phalloides*), Destroying Angel (*A. verna—virosa* complex), Galerinas (*Galerina* spp.), and some Corts (*Cortinarius* spp.), as well as in some poisonous bluegreen algae. European Mistletoe (*Viscum album*) and Akee seeds (*Blighia sapida*) are two other plants containing toxic polypeptides.

Poisonous proteins are called toxalbumins or phytotoxins. Most are either lectins or enzyme inhibitors. A lectin called ricin is said to be the most toxic naturally occurring compound. Found in the Castor Bean (*Ricinus communis*), it inhibits protein synthesis in the cell wall and agglutinates red blood cells. Other toxic lectins include: abrin, found in Rosary Pea (*Abrus precatorius*); robin, found in Black Locust (*Robinia pseudo-acacia*); and a lectin found in Pokeweed (*Phytolacca americana*), which is a mitogen, affecting the activity of white blood cells.

Thiaminase, the thiamine-destroying enzyme found in Bracken Fern (*Pteridium aquilinum*), is another example of a toxic protein.

In addition to these toxic proteins, others are responsible for the specific allergic reactions some people have towards certain types of plants. In these cases, the intensity of the reaction depends not on the quantity of the toxic protein but on the sensitivity of the affected person.

Resins and Volatile Oils: Resins and volatile, or essential, oils are complex and diverse compounds widely produced by plants. Many are said to be derived from terpenes, which are hydrocarbons having multiples of the molecular formula C_5H_8 and are believed to be the most numerous organic compounds in plants. Resins and volatile oils often occur together in mixtures called oleoresins. Turpentine, for example, is an oleoresin obtained from sapwood ducts of certain pines (*Pinus* spp.).

Mayapple (*Podophyllum peltatum*) contains a complex of toxic compounds known in commerce as podophyllum resin. These are bitter, irritant, and strongly purgative. Podophyllotoxin and its derivatives have medicinal potential in cancer therapy. Only the ripe fruit of Mayapple is edible.

Spurges, including Crown-of-Thorns, Poinsettia, and various other ornamental species (*Euphorbia* spp.), contain an irritant compound, phorbol, as a major toxic component of their latex. Many members of the heather family (Ericaceae), including Rhododendrons and Azaleas (*Rhododendron* spp.), Mountain and Swamp Laurels (*Kalmia* spp.), and Pieris (*Pieris japonica*), contain terpenes generally known as andromedotoxins, which affect the heart and circulatory system and are potentially deadly. Other toxic terpenes are found in Daphne (*Daphne mezereum* and related spp.), Iris (*Iris* spp.), Marijuana (*Cannabis sativa*), and Chinaberry Tree (*Melia azedarach*).

Ester derivatives of some resins, including phorbol from the Spurges and their relatives and mezerein from *Daphne,* are known to present another hazard in addition to being directly poisonous. Termed "cocarcinogens," they act as potent cancer-causing agents when applied after a low-dosage application of a carcinogen. The initial carcinogen may "initiate" a "latent tumor cell," which may remain normal in appearance for a long period. Then, after the second substance, the "cocarcinogen" is applied, the cell rapidly becomes cancerous. The "cocarcinogen" however, does not cause malignant growth itself, even when applied repeatedly. Details of this phenomenon are given by Kinghorn (1979).

Volatile oils are almost all mixtures of compounds, usually a liquid and one or more solid components. They are strongly scented and, as their name suggests, they evaporate rapidly when exposed to air. They occur in various plant tissues, depending on the plant family. They are the compounds responsible for many scents of flowers, herbs, and spices. Some, such as bitter almond oil and mustard oil, are produced from glycosides, and hence are often classed together with glycosides.

Many volatile oils are fragrant and pleasant flavorings of our foods and beverages. In larger amounts, however, they can be harmful.

Some are irritant, and some may be tumor-inducing. Thujone is an volatile oil occurring in various Cedars, Cypresses, and Junipers (*Thuja* spp., *Chamaecyparis* spp., *Juniperus* spp.), and also found in various species of the aster, or composite family (Asteraceae, or Compositae), including Wormwoods (*Artemisia* spp.), Tansy (*Tanacetum vulgare*), and Yarrow (*Achillea millefolium*). If used repeatedly, thujone can cause serious personality changes, and in large quantities it can produce convulsions and brain cortex lesions. Because of these problems the liqueur known as absinthe, flavored with *Artemisia absinthium*, was banned in France after being widely used there and causing harm to many people. Other potentially harmful volatile oils include: umbellulone, from the leaves of California Bay Laurel (*Umbellularia californica*); asarone, from the roots of Sweet Flag (*Acorus calamus*) and Wild Ginger (*Asarum* spp.); menthol, from Peppermint oil (*Mentha piperita*); camphor, from the Camphor Tree (*Cinnamomum camphora*) and synthetic sources; myristicin, from Nutmeg and Mace (*Myristica fragrans*); and safrole, from Sassafras (*Sassafras albidum*) and other spicy flavorings (see Hall 1973).

Phototoxins: Some plants contain chemical substances that can make the skin extremely sensitive to ultraviolet radiation in sunlight. These substances are known as phototoxins, or photosensitive agents, and they are a diverse group of compounds. Some can cause cell damage simply on contact with the skin. These include the furanocoumarins in various plants in the celery, or umbel family (Apiaceae, or Umbelliferae) and thiophene compounds in Marigold (*Tagetes* spp.) and other members of the aster, or composite family (Asteraceae, or Compositae). Other phototoxic substances travel to the skin after being eaten and absorbed into the blood stream. In this category are phototoxins in St. John's-Wort (*Hypericum perforatum*), Buckwheat (*Fagopyrum esculentum*), Knotweeds (*Polygonum* spp.) and Lamb's Quarters (*Chenopodium album*). A secondary type of photosensitization, occurring in livestock, results from any type of damage to the liver which prevents the normal excretion of a common pigment, phylloerythrin, produced in the digestive tract from chlorophyll. The phylloerythrin then accumulates in the blood and tissues and causes skin reactions in sunlight.

Symptoms of photosensitization resulting from cell damage to the skin include itchiness, redness, heat, swelling, and blistering that may last for many days, weeks, or even months. Excessive pigmentation, or hyperpigmentation, of the skin in the affected area may remain for a year or more. Photodermatitis is sometimes difficult to distinguish from skin reactions caused by Poison Ivy and its relatives, which are allergens. The condition, being activated by ultraviolet light, occurs only if the skin of the person or animal is exposed to sunlight (or to artificial uv light) after exposure to a phototoxic substance. Light-skinned people and animals are more likely to be affected by phototoxins than dark-skinned.

It seems somewhat paradoxical, but some phototoxins, such as the furanocoumarins psoralen and 8-methoxypsoralen, have been used recently with excellent initial results in treating psoriasis, a debilitating skin disease affecting about 2 % of the world's population. The phototoxins, given orally and followed with exposure to certain wavelengths of ultraviolet radiation, prevent the excessive cell division of skin cells that is characteristic of psoriasis. For further information on phototoxins, the reader is referred to Kingsbury (1964), Scheel (1973), Towers (1979), and Fuller and McClintock (1986).

Cancer-causing Plant Substances: Many plant substances are known to increase the likelihood of the growth of abnormal or cancerous cells in laboratory animals such as rats and mice. Humans can be expected to have similar reactions to these substances, although in most cases their hazards can only be assumed. A review of naturally occurring, tumor-inducing substances is given by Miller (1973) in a volume produced by the National Academy of Sciences (1973). Although cancer-causing plants must certainly be considered toxic, it is not within the scope of this book to treat the various types in detail. So little is known as yet about many types of cancer. In many cases the reasons why some people are affected by cancer, and others living under similar circumstances are not, are not fully understood. We do know that some substances statistically increase the risks of contracting cancer, and for this reason certain plants and foods definitely should be avoided.

Fungi, especially molds, are known to produce many tumor-inducing substances. For example, aflatoxins, potent cancer-causing substances acting mainly on the liver but also on other tissues, are produced by certain molds (mainly *Aspergillus flavus*) growing on Peanuts. It has been suggested that there is a direct link between high rates of liver cancer and consumption of mold-contaminated Peanuts in various parts of the world. The grain-contaminating fungus, Ergot (*Claviceps* spp.), contains many alkaloids and other physiologically active substances, and has also been found to induce tumors in laboratory animals.

Pyrrolizidine alkaloids, occurring in Ragworts (*Senecio* spp.), Heliotropes (*Heliotropium* spp.), and Rattlebox (*Crotolaria* spp.), are potent liver and lung toxins and are also suspected of being carcinogenic under some circumstances.

Safrole, a volatile oil, is present in several different flavorings and spices, especially the root bark of Sassafras (*Sassafras albidum*), a favorite wild tea. With prolonged ingestion of large concentrations, it has caused the growth of liver tumors in laboratory rats; because of this, safrole is no longer used as a flavoring agent in root beers.

Asarone, chemically related to safrole and occurring in Sweet Flag (*Acorus calamus*), has also been implicated as a tumor-causing agent. In both cases, however, there is no direct evidence of these substances causing cancer in humans.

Cycads, or seed-ferns (*Cycas* spp.) are commonly grown as house and greenhouse ornamentals, as well as outdoors in warmer areas. They contain a series of toxic glycosides, the principal one being cycasin. Their seeds, produced on large, cone-like structures, have been found to be highly carcinogenic when fed to rats and other animals. Cycads have been used as an emergency and staple food by native peoples in the tropics, but their use is not recommended under any circumstances.

Bracken Fern (*Pteridium aquilinum*), whose green fiddleheads and rhizomes have been widely eaten by certain ethnic groups, have been found to contain several cancer-causing substances, as well as other toxins. It has been suggested that the high incidence of stomach cancer in Japan, New Zealand, and the United States may be partially linked to eating Bracken shoots.

Numerous other substances have been linked with cancers in laboratory studies and surveys, but how closely laboratory conditions simulate ordinary living conditions for people is still the subject of much speculation.

CHAPTER II

POISONOUS MUSHROOMS

WHAT ARE MUSHROOMS?

Mushrooms are fleshy, spore-bearing structures which are easily visible. They are familiar to adults and children alike from their most characteristic shape, a rounded cap on a central stalk. Mushrooms belong to a large, complex group of organisms called fungi, all of which lack chlorophyll, the green substance that enables green plants to manufacture their own food through photosynthesis. Lacking chlorophyll, fungi obtain their food from decaying plant and animal remains, or, in the case of parasitic fungi, from living plants or animals. Mushrooms are only the reproductive part of the organism—the "fruit"; the main part is a seldom-seen mass of tiny thread-like growths, or hyphae, called the mycelium. Ever present but usually inconspicuous, mycelia penetrate soil, bark, and wood. Often a special symbiotic, mutually beneficial relationship develops between fungal mycelia and the roots of particular higher plants, such as some coniferous trees. The fungi in such associations are called mycorrhizae. Some fungi also develop special, close relationships with particular kinds of algae, and the resulting two-species organisms are called lichens.

Mushrooms make up only a small fraction—about 5,000 to 10,000—of the total number of fungi, which includes an estimated 200,000 or more species, the majority being inconspicuous or microscopic. Although many fungi are edible or useful to people in some way, many are potentially harmful. Of these, the smaller, less visible types are discussed elsewhere (p. 63). Poisonous mushrooms, or "toadstools," are treated here as a distinct group. Only general descriptions are provided. Exact identifications of many species must be carried out at a microscopic level, and usually require the skills of a fungus expert, or mycologist.

MUSHROOM POISONING

There are many safe, edible, even choice, species of wild mushrooms in North America, but some are poisonous to varying degrees. A few are highly toxic, even fatal, if eaten. The actual number of edible or poisonous species is impossible to determine because most mushrooms are not collected for food, and most have not been analyzed for toxins.

In recent years the number of cases of mushroom poisoning in North America has increased considerably, and poison control centers everywhere are called on to cope with this problem, especially in the fall. New information on the identification and distribution of toxic mushrooms, and on the treatment of mushroom poisoning, is appearing all the time. In the autumn of 1988, for example, a British Columbia man was fatally poisoned from eating mushrooms. *Lepiota subincarnata,* a species previously not reported from Canada, was linked to this death, according to Dr. Scott Redhead of the Biosystematics Research Centre in Ottawa. Kidney dialysis and liver transplants are now being used to save the lives of some mushroom poisoning victims who would have died from more conventional treatment.

Typical symptoms of mushroom poisoning include nausea, diarrhea, cramps, vomiting, and in some cases, drowsiness, hallucinations, or even coma. Depending upon the toxins involved different combinations of these symptoms may be exhibited. There is a great deal of variation among individuals in their response to the less toxic mushrooms. We know that some people react adversely to species that are harmless to most, and some may experience unpleasant effects from a type of mushroom they had eaten on previous occasions without any bad reaction. People can be allergic to certain mushrooms just as they can to other types of foods such as peanuts or wheat products. People sick with infections or flu may react to eating mushrooms. Some become upset after eating mushrooms out of fear alone, from nervousness about eating a species they have never tried before. In many cases, gastrointestinal upset caused by mushrooms is simply overindulgence, especially if the dish is cooked with butter, sour cream, or bacon. However, with the more serious toxins symptoms vary little and can be fatal.

The poisonous qualities of mushrooms can vary with state of maturity, geographical location, and other environmental and genetic factors. Some species are considered poisonous in one region, and edible and completely safe in another. Furthermore, any edible species can become harmful if infected by toxic organisms such as some molds, or if sprayed with herbicides or pesticides. If growing alongside busy roads, they can absorb poisonous metals such as mercury, lead, or cadmium.

POINTS TO REMEMBER FOR MUSHROOM GATHERERS

1. There are no rules of thumb or simple tests for determining if a mushroom is edible or poisonous.
2. Before eating any wild mushroom, be absolutely certain of its identification and edibility. Collect only firm, fresh mushrooms without insects or worms. Store mushrooms in paper bags, or in waxed paper, in a cool place. Unless otherwise advised, cook all wild mushrooms; do not eat them raw.
3. When eating for the first time a wild mushroom which has been identified as an edible species, consume only a small portion, and do not drink any liquor. Do not eat more than one kind of mushroom at a time (eating mixtures complicates identification of stomach contents). If no side effects occur within 48 hours, try a slightly larger portion the next time, but never eat a very large quantity, no matter how often a particular species has been eaten.
4. When trying a new type of mushroom, save one or two whole specimens to provide positive identification and proper treatment should any ill effects occur.
5. Do not eat any *Amanita* species (even though some are edible), and be careful in identifying *Amanita* look-alikes. It is a good idea to avoid any mushroom with warty spots on the cap, white gills, a ring on the stem, and/or a globular or cup-like structure at the base of the stem.
6. Avoid little brown mushrooms, and large brownish mushrooms, especially those with pinkish, brownish, purple-brown, or blackish gills; these can be confused with the highly toxic Galerinas (*Galerina* spp.) or with some of the toxic *Cortinarius* spp.
7. As with gilled mushrooms, do not eat any boletes (with pores rather than gills beneath the cap) unless identification is certain. Above all, avoid species in which the pore surface is red or orange and those in which the cut flesh turns quickly to blue.
8. Should poisoning occur from eating mushrooms, prompt treatment can make the difference between recovery and death. Induce vomiting, consult a physician, and take the patient to a hospital immediately. Bring a sample of the mushroom, ideally whole and uncooked, but even a processed and cooked sample is far better than nothing. Save any stomach contents from vomiting to help in identification. The physician should request immediate identification of the mushroom through the authorities known to the poison center, or through an accredited mycologist. Often, microscopic examination of the mushroom is required to make a positive identification. A well-written, well-illustrated mushroom identification book, such as *The Audubon Society Field Guide to North*

American Mushrooms by Lincoff (1981), is extremely helpful. A thorough, technical treatment of toxic mushrooms is provided in *Poisonous Mushrooms of the Northern United States and Canada* by Ammirati, Traquair, and Horgen (1985), and in *Toxic and Hallucinogenic Mushrooms. A Handbook for Physicians and Mushroom Hunters* by Lincoff and Mitchel (1977). Refer also to *Mushroom Poisoning: Diagnosis and Treatment,* edited by Rumack and Salzman (1978), and the Canadian Pharmaceutical Association's *Poison Management Manual* (1984), under Mushroom Poisoning.

This book cannot replace the advice and assistance of qualified medical personnel.
In all cases of suspected poisoning by plants, or any other substance, immediate qualified medical advice and assistance should be sought.

TYPES OF POISONOUS MUSHROOMS

Poisonous mushrooms are often classified by the type of poisons they contain. Eight types of toxins are generally recognized. The most important poisonous species are included within Types I-VII. Although it is not possible to deal here with all known poisonous species, selected examples from all eight groups of poisonous mushrooms are discussed in this book.

TYPE I
Poisonous Substance(s): amanitins (amatoxins) and phalloidins (phallotoxins) (cyclic peptides)
Species: *Amanita phalloides* (Death Cap); *A. verna—virosa* complex (Destroying Angel); *Galerina autumnalis* (Autumn Galerina) and various other *Galerina* species; some *Lepiota* species, including *L. subincarnata, L. helveola,* and *L. josserandii; Conocybe filaris* (Deadly Conocybe)
Symptoms: delayed for 6–24 hours after ingestion; abdominal pains, nausea, vomiting, diarrhea lasting 1 or more days, often followed by short remission; then recurring pain, liver and kidney disfunction, convulsions, coma, and often death; recovery, with proper treatment, can occur in 1–2 weeks, but victim can have permanent liver and kidney damage.
Treatment: see under *Amanita phalloides* (p. 39) and *Galerina autumnalis* (p. 49)

TYPE II
Poisonous Substance(s): orellanins (cyclic peptides)

Species: *Cortinarius gentilis* (Deadly Cortinarius), *C. orellanus* (Poznan Cortinarius) and various other *Cortinarius* species

Symptoms: delayed 3–14 days after ingestion; acute or chronic kidney failure which can result in death; recovery with treatment can take as long as 6 months.

Treatment: see under *Cortinarius* spp. (p. 47)

TYPE III
Poisonous Substance(s): gyromitrin (monomethyl hydrazine—MMH)

Species: *Gyromitra esculenta* (False Morel) and various other *Gyromitra* species

Symptoms: delayed usually 6–12 hours after ingestion; bloated feeling, nausea, vomiting, watery or bloody diarrhea, abdominal pains, muscle cramps, faintness, loss of coordination, and sometimes convulsions, coma, and death; with treatment, recovery can occur within hours.

Treatment: see under *Gyromitra esculenta* (p. 50)

TYPE IV
Poisonous Substance(s): muscarine

Species: *Clitocybe dealbata* (Sweat-causing Clitocybe) and other *Clitocybe* species; *Inocybe geophylla* (White Inocybe) and most other *Inocybe* species; also present, but not the dominant toxin, in *Omphalotus* spp. (Jack O'Lantern) and *Amanita muscaria* (Fly Agaric) and its relatives.

Symptoms: profuse perspiration, salivation, tears, blurred vision, abdominal cramps, watery diarrhea, constriction of pupils, drop in blood pressure, slow pulse; deaths relatively frequent from some *Inocybe* spp.; occasional death from other types in people with pre-existing illness; recovery usually occurs within 6–24 hours.

Treatment: see under *Inocybe geophylla* (p. 52)

TYPE V
Poisonous Substance(s): ibotenic acid and muscimol (mycoatropines)

Species: *Amanita pantherina* (Panther Agaric), *A. muscaria* (Fly Agaric), and various other *Amanita* species

Symptoms: occur 30 minutes to 2 hours after ingestion; dizziness, lack of coordination, delusions, staggering, delirium, muscular cramps, hyperactivity, followed by deep sleep; recovery usually within 4–24 hours; there are cases on record of only one bite causing a victim to go berserk, or crazy.

Treatment: see under *Amanita pantherina* (p. 37)

TYPE VI

Poisonous Substance(s): psilocin, psilocybin (indole derivatives)

Species: *Psilocybe semilanceata* (Liberty Cap), *P. cubensis* (Common Large Psilocybe), and various other *Psilocybe* species ("Magic Mushrooms"); *Conocybe smithii* (Bog Conocybe), *Gymnopilus spectabilis* (Big Laughing Gymnopilus), *Paneolus subbalteatus* (Girdled Paneolus)

Symptoms: usually occur within 30–60 minutes; mood changes; laughter, compulsive movements, weakness of muscles, drowsiness, visions, sleep; recovery usually within 6 hours.

Treatment: see under *Psilocybe semilanceata* (p. 56)

TYPE VII

Poisonous Substance(s): coprine (antabuse-like or disulfiram-like compounds)

Species: *Coprinus atramentarius* (Alcohol Inky-cap) and some other *Coprinus* species, *Clitocybe clavipes* (Fat-footed Clitocybe)

Symptoms: occur 30 minutes or so after drinking alcohol, as long as 5 days after eating mushrooms; flushed face, distension of neck veins, swelling and tingling of hands, metallic taste in mouth, palpitations, low blood pressure, nausea, vomiting, perspiration; recovery usually within 2–4 hours. Symptoms also occur if alcohol is ingested just prior to eating the mushrooms.

Treatment: see under *Coprinus atramentarius* (p. 45)

TYPE VIII

Poisonous Substance(s): diverse, often unknown, gastrointestinal irritants

Species (some examples): *Agaricus meleagris* (Western Flat-topped Agaricus) and some other *Agaricus* species; *Armillariella mellea* (Honey Mushroom); *Boletus subvelutipes* (Red-mouth Bolete); *Chlorophyllum molybdites* (Green-spored Lepiota); *Entoloma sinuatum* (Lead Poisoner) and some other *Entoloma* species; *Hebeloma crustuliniforme* (Poison Pie); *Lactarius torminosus* (Pink-fringed Lactarius) and some other *Lactarius* species; some *Lepiota* species; *Omphalotus* spp. (Jack O'Lantern); *Russula emetica* (Emetic Russula); *Scleroderma* spp. (Poison Puffball); *Tricholoma pardinum* (Dirty Tricholoma); *Tricholoma pessundatum* (Red-brown Tricholoma)

Symptoms: usually occur within 30 minutes to 3 hours after ingestion; mild to severe nausea, vomiting, diarrhea, abdominal pain; recovery is normally complete and fairly rapid, usually from 1 hour to 2 days. Sometimes (as in the case of *Chlorophyllum molybdites*) poisoning is more serious, and recovery may take several weeks.

Treatment: general supportive therapy for symptoms; empty

stomach; follow with activated charcoal and a saline cathartic; maintain fluid and electrolyte balance; antacids may provide relief; confirm identification of mushroom.

Much remains to be learned about mushroom poisoning, and many dangerous mushroom toxins are still to be isolated and analyzed. The mushrooms described in this book are listed in alphabetical order of their scientific names, since the common names are often quite variable, and are frequently derived from the scientific names.

THE *AMANITA* GROUP

Amanitas are the most important group of toxic mushrooms, with three species in the genus being responsible for most mushroom fatalities, and others causing serious illness or occasionally death if not treated. Most of the Amanitas are conspicuous, large, and strikingly beautiful. There are no easily distinguished traits—color, taste, odor or any other characters—that separate them from other genera. White Amanitas are often mistaken for edible *Agaricus* or *Volvariella* species.

Amanitas do share the following features that will help to identify them (see diagram):

1. All Amanitas start to develop as round or oval structures known as "buttons".
2. Both a universal and partial veil are present in the button stage of most species (see diagram) and can be seen if the button is cut lengthwise.
3. As the stem of the button elongates both veils break. Remnants of the outer veil can often been seen on the cap as warty patches of tissue and at the base of the stem as a cup-like structure, the volva. When the inner, or partial, veil breaks it remains on the stem as a ring. The ring is sometimes poorly formed or very delicate and may disappear altogether. In some species (e.g., Fly Agaric, *Amanita muscaria*) the volva is not free from the stem but remains as a series of ridges running around the base of the stem.
4. All Amanitas have white or pale creamy gills, which are free from the stem or just reach it.
5. All Amanitas have white spores and spore prints.

There are actually a number of edible *Amanita* species in North America, and some have been eaten in Europe for centuries. However, there are more species in North America than in Europe and many are still poorly known. Considering the fact that the edible types closely resemble some of the very toxic kinds, eating *any* Amanitas is strongly discouraged.

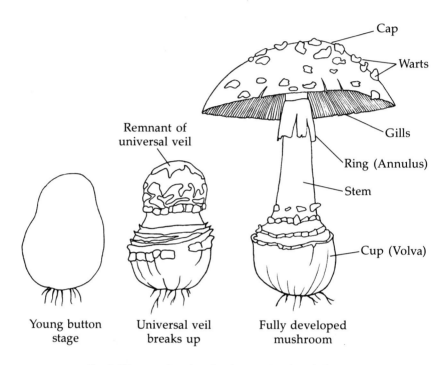

Fig. 5. Diagram showing development and typical parts of *Amanita* mushroom.

Fly Agaric, or Fly Mushroom
(Amanita Family)

Amanita muscaria
(Amanitaceae)

Fig. 6. Fly Agaric (*Amanita muscaria*), orange eastern form. MARY W. FERGUSON

QUICK CHECK Scarlet red, orange, or yellowish, medium to large-sized mushroom of forests, clearings, or lawns near trees or shrubs; cap covered with whitish, warty spots; stem and gills white; several rings or ridges around the swollen base of the stem. Hallucinogenic, with symptoms resembling alcohol intoxication; may cause some individuals to go berserk, or cause temporary coma; seldom fatal. First symptoms appear within an hour or so.

DESCRIPTION *Cap* 5–30 cm (2–12 in.) across, scarlet-red mainly in western North America, orange to yellow-orange mostly in eastern North America, fading to creamy yellow; sticky when wet; usually covered with whitish warts or patches (may disappear with age); rounded when young, becoming flat or saucer-like with age; *Flesh* firm, white or creamy, without distinctive odor; *Gills* whitish, crowded, extending to stem but free from it; *Stem* whitish, smooth or covered with silky hairs; up to 20 cm (8 in.) long and 2.5 cm (1 in.) thick, enlarged at base,

Fig. 7. Fly Agaric, red western form. IAN G. FORBES

Fig. 8. Fly Agaric, showing variations of "wart" patterns and coloration, eastern form. BIOSYSTEMATICS RESEARCH CENTRE, AGRICULTURE CANADA

becoming bulbous; *Ring* white, membranous, hanging, and conspicuous; *Cup (Volva)* white, in the form of a few concentric rings above the swollen stem base; occasionally with a single ring and forming a very shallow cup; *Spore print* white.

OCCURRENCE Singly, or sometimes grouped, in rings or arcs on the ground in hardwood, coniferous, or mixed forests, as well as forest openings, grassy areas, and lawns under trees. Found throughout temperate zone of Northern Hemisphere; common in North America from spring through fall. Yellow and orange forms predominate in northern and eastern North America, and red on the Pacific Coast.

TOXICITY Deaths from Fly Agaric poisoning are rare and its reputation as a deadly poisonous mushroom may be over-rated, but the toxins and their effects on individuals vary greatly. Fly Agaric contains the hallucinogens ibotenic acid and muscimol and the toxin muscarine in usually insignificant amounts (estimated at 0.0025 percent dry weight basis), varying somewhat from one location to another. Contrary to some reports, Fly Agaric does not contain atropine.

The mushroom is variable in its effects; eating just one mouthful can cause a serious reaction in some people. The first symptoms, occurring in 30–60 minutes, resemble those of alcohol intoxication, with drowsiness and dizziness. This stage is usually followed by confusion, muscular spasms, delirium, and visual disturbance lasting a few hours and generally succeeded by drowsiness and deep sleep. Sometimes the victim becomes comatose. Recovery is usually quite rapid, within 6–24 hours, depending on the amount ingested and the state of health of the patient. Occasionally the mushroom causes very violent reactions; one man in British Columbia went berserk from ingesting a single bite. Some researchers have characterized the action of ibotenic acid and muscimol on neural transmission as resembling that of the drug LSD.

TREATMENT The action of ibotenic acid and muscimol is similar to that of atropine; therefore atropine *should not be* administered. The small amount of muscarine is not likely to produce symptoms of muscarine poisoning. Induce vomiting, or perform gastric lavage; general supportive measures; control marked excitation with i.v. diazepam; keep patient quiet and reassured.

NOTES A major problem with Fly Agaric is that it can easily be misidentified. It is so variable in form, and there are many "look-alike" types. These vary greatly in relative toxicity and some, such as Panther Agaric (*Amanita pantherina*), can be deadly.

Fly Agaric is probably the best known mushroom in the world. It is strikingly attractive and is illustrated in many children's books, as well as being used in a variety of graphic designs. Its common name, Fly Agaric, comes from its long-standing use as an insecticide against houseflies. However, Adam Szczawinski recalls this use from his childhood days in Poland, and remembers that many of the flies were not actually killed from eating it, but only intoxicated for a few hours. Fly Agaric has a long history of use as an intoxicant and hallucinogen by peoples of Europe and Asia, particularly Northern Siberia and the Kamchatka Peninsula. People in these last regions were known to drink the urine of individuals intoxicated by the mushroom, becoming similarly intoxicated thereby. Apparently most of the hallucinogenic compounds pass through the body intact. Fly Agaric is the subject of several interesting books, including *"Soma": Divine Mushroom of Immortality* by R. Wasson (1968) and *The Sacred Mushroom and the Cross* by J. M. Allegro (1970). Its history as an hallucinogen is described by Emboden (1979).

Panther Agaric, Panther Amanita, or Panther Cap
(Amanita Family)

Amanita pantherina
(Amanitaceae)

Fig. 9. Panther Agaric (*Amanita pantherina*), young stage, showing cup, or volva. R. & N. TURNER

QUICK CHECK Medium-sized mushroom of open woods or wooded lawns; cap pale to dark brown, usually with warty white spots; stem and gills white; ring around stem; volva, or cup, at base of stem.

Highly poisonous, occasionally fatal. Symptoms usually appear within an hour.

DESCRIPTION *Cap* 5–15 cm (2–6 in.) across, occasionally larger, rounded at first, becoming flat in age, viscid (sticky) when moist, pale tan to chocolate-brown, covered with whitish, pointed warts; *Flesh* firm, whitish without distinctive odor or taste; *Gills* white, closely spaced, free from stem; *Stem* up to 12 cm (5 in.) long and 2.5 cm (1 in.) thick, white, silky, shiny above the ring and somewhat hairy below the ring, enlarged and bulb-like at the base, arising from cup; *Ring* single, membranous, cottony above and fibrillose (stringy) below; *Cup (Volva)* adhering to stem forming a narrow, free roll or collar where the stem enlarges; *Spore print* white.

Fig. 10. Panther Agaric, mature stage.
MARY W. FERGUSON

OCCURRENCE Throughout temperate northern hemisphere; common in northwestern North America and Rocky Mountains. Grows singly or in groups, occasionally in rings or arcs, on the ground in open coniferous or mixed woods; also on lawns and in gardens and parks in the vicinity of mature coniferous trees, especially Douglas-fir. Found from early spring to late fall, occasionally even in winter when weather is mild.

TOXICITY Panther Agaric is seriously toxic, and occasionally fatal. The main poisonous compounds are ibotenic acid, muscimol, and pantherin, all of which may vary in concentration. There are clinically insignificant amounts of muscarine (0.0025 percent dry weight basis) and related compounds.

Symptoms appear within 15–60 minutes after eating. A feeling of drowsiness is followed by a state resembling alcohol intoxication. Then a state of confusion sets in, accompanied by muscular spasms, delirium and disturbance of vision lasting a few hours. Vomiting seldom occurs. Drowsiness and sleep follow and recovery is quite rapid, usually within 24 hours, with little or no memory of the event.

Fatalities, mostly in children, sometimes occur. Puppies and other young pets may also be fatally poisoned.

TREATMENT The action of ibotenic acid and muscimol is similar to that of atropine. Consequently, atropine *should not be* prescribed or administered. The minute amount of muscarine present is not likely to produce muscarinic symptoms. Induce vomiting, or perform gastric lavage; general supportive measures; maintain fluid and electrolyte balance. Nikethamide may be given if patient is markedly depressed, and i.v. diazepam for marked excitation. Keep patient quiet and provide reassurance.

NOTES This common, highly toxic *Amanita* has been responsible for poisoning a number of children. It frequently occurs on lawns and in woods in populated areas, making the danger of poisoning great. The mushroom is named from its panther-like spots, or warts, on the cap. In the Pacific Northwest it intergrades with the closely related but usually less poisonous *Amanita gemmata*, giving rise to a series of confusing forms with caps ranging from light tan to dark brown and containing widely variable amounts of toxins.

Death Cap
(Amanita Family)

Amanita phalloides
(Amanitaceae)

Fig. 11. Death Cap (*Amanita phalloides*).
KIT SCATES BARNHART

QUICK CHECK Attractive, medium-sized mushroom of oak, beech, pine, or mixed woods and plantings; cap smooth, yellowish green to greenish brown; stem and gills white; stem bulbous-based, with sac-like cup, or volva, at ground level, and large, skirt-like ring. Highly toxic, often fatal. First symptoms appear in 6–24 hours, often followed by period of apparent improvement, then by symptoms of liver failure after 4 or more days.

DESCRIPTION *Cap* 5–15 cm (2–6 in.) across, yellowish green to greenish brown, rarely very pale or nearly white, and faintly streaked radially with darker fibrils near the center; convex or flat when fully expanded; slightly sticky; easily peeled; without warty spots, or occasionally with only a few dried traces of white veil; *Flesh* firm, white to light green just below cap; odor becoming foul; *Gills* white, close together but well separated and free from stem; *Stem* up to 15 cm (6 in.) long and nearly 2 cm (0.8 in.) thick, enlarged toward the base, white, smooth, solid when young, but hollow when mature; *Ring* thin, membranous, persisting and hanging skirt-like on stem; *Cup (Volva)* whitish, sac-like, irregularly lobed at the base of the bulbous stem; persistent, often buried in ground; *Spore print* white.

OCCURRENCE On ground in oak, beech, pine, or mixed woods. Infrequent in North America, sometimes occurring under plantations of European trees. Known from Massachusetts to Virginia and west to Ohio in the East, and from Washington to California in the West, where it has been found with increasing frequency in recent years. Common in Central Europe, and less abundant in Asia, Japan, China, and New Zealand.

TOXICITY This mushroom is responsible for more deaths than any of the other species. It contains two closely related groups of toxins, the amatoxins, which are cyclic octapeptides, and the phallotoxins, which are cyclic heptapeptides. There are six known amatoxins, including amanitin, and five phallotoxins, including phalloidin. The relative toxicity of amatoxins and phallotoxins is still unclear. Although it is generally accepted that the amatoxins (mainly alpha-amanitin), and not phallotoxins, are responsible for the lethality of *Amanita phalloides* and related species, some recent experimental data contradict this hypothesis (Floersheim 1987).

Poisoning symptoms occur in three distinct stages. The first, developing within 6–24 (usually 10–14) hours, include: dry mouth, nausea, vomiting, sharp abdominal pains, diarrhea (often with blood and mucus), and shock. The longer these initial symptoms are delayed, the better an individual's chances for survival. After about 24 hours, these symptoms may be followed by a period of false recovery lasting up to 4 days. Then comes the most serious stage of intoxication, with recurrence of abdominal pain and symptoms of liver and kidney failure frequently leading to death in 7–10 days due mainly to liver deterioration and subsequent failure, blood coagulation deficiency, and ultimately, hepatic coma and heart and kidney failure.

About 30 g (half a fresh mushroom) can be fatal to an adult. The larger the quantity consumed, the greater the risk of fatality. The general mortality rate for Death Cap poisoning ranges between 20% and 30%. Children, however, are at much greater risk, with a mortality rate of over 50%. This is possibly due to the fact that they often eat similar quantities of mushrooms as adults, but because of their smaller size, absorb a larger dose of the toxins in proportion to their bodyweight.

This mushroom is also toxic to dogs and some other animals, but cases of poisoning are rather rare. Certain animals, including rabbits and gray squirrels, are not affected by oral ingestion of the mushroom.

TREATMENT Treatment must be started as soon as possible, to prevent irreversible damage to liver and kidneys. Correct identification of the mushroom responsible is vital. Hospitalize immediately; induce vomiting; perform gastric lavage as soon as possible up to 36 hours after ingestion to prevent further absorption of toxins; follow with activated charcoal to absorb residual toxins in the stomach and intravenous fluids to correct fluid and electrolyte imbalance from vomiting and diarrhea. *Do not* use antidiarrheal or antispasmodic agents to control digestive upset.

Monitor blood pressure, arterial blood gases, serum electrolytes, glucose, liver and kidney function, heart beat, and thromboplastin time (for blood coagulation deficiency). This last factor is the most reliable indicator for the severity of the poisoning. Maintain urine output with diuretics if necessary. SGOT (serum glutamic oxaloacetic transaminase) and SGPT (serum glutamic pyruvic transaminase) enzyme tests should be performed promptly and followed daily, to monitor liver and renal damage.

Many treatments have been used against poisoning by Death Cap but, according to Floersheim (1987), few have ultimately been found to be effective. In a review of various treatments, Floersheim concludes that only benzylpenicillin (Penicillin G), administered in doses of 300,000 to 1,000,000 U/kg/day, has been shown in both experimental and clinical studies to be a helpful antidote. Another substance which seems very promising is silibinin, a compound derived from Milk Thistle (*Silibum marianum*). It is thought to inhibit the penetration of amatoxins into liver cells, and although its therapeutic value has not been statistically proven it has an impressive clinical record. It should be administered early (within 48 hours if possible) in a dose of 20 to 50 mg/kg/day. Silibinin is not always readily available, and the United States Food and Drug Administration has only approved oral use of Milk Thistle extract, not intravenous application.

Several other treatments have been recommended at various times, but no antidote to date is totally effective, and too many different treatments applied at the same time may compromise the patient's chances of recovery. In general, treatment should consist of supportive therapy to counter the effect of severe liver and kidney damage induced by the toxins. In Europe, kidney dialysis machines are frequently used, with some success, to treat *Amanita phalloides* poisoning. Liver transplants have also been necessary in some cases.

NOTES In the fall of 1988, five people were seriously poisoned after they confused *Amanita phalloides* buttons growing in a park in Vancouver, Washington with Paddy Straw Mushrooms (*Volvariella volvacea*), a popular Oriental mushroom sold commercially all over Asia. After three and a half days, the victims' livers were irreversibly damaged, and four out of the five required liver transplants.

Amanita phalloides has been occasionally confused with *A. brunnescens*, another poisonous species, common in hardwood and mixed forests of eastern North America. Its basal bulb is cleft, or split, and its cap is viscid brown with white, cottony warts. Species of *Russula* with greenish caps can also be mistaken for *Amanita phalloides*, but the Russulas lack the basal volva, or cup, and skirt-like ring.

Destroying Angel, Death Angel, or Spring Amanita
(Amanita Family)

Amanita verna, A. virosa, and
    closely related species
(Amanitaceae)

Fig. 12. Destroying Angel (*Amanita verna*).
MARY W. FERGUSON

QUICK CHECK Pure white mushrooms of forests or clearings; cap smooth, sticky when wet; sac-like volva (cup) ensheathing swollen stem base; large, skirt-like ring around stem. Highly toxic, often fatal. First symptoms appear in 6–24 hours, followed by period of apparent improvement, then by symptoms of liver and kidney failure after 4 or more days.

DESCRIPTION *Cap* pure white, 3–10 cm (1–4 in.) across, rounded, initially egg-shaped, becoming flat; smooth, sticky when wet; *Flesh* firm, white, sometimes with potato-like odor; *Gills* white, crowded, free from stem; *Stem* up to 25 cm (10 in.) long and 2.5 cm (1 in.) thick, smooth (in *A. verna*) to cottony or shaggy (in *A. virosa*), white, enlarged at base; *Ring* large, white, skirt-like, flaring downwards and outwards; *Cup (Volva)* white, persistent, sac-like, free from stem for some distance above the swollen base; *Spore print* white.

OCCURRENCE On ground, solitary or in small groups, in hardwood, mixed, or occasionally coniferous forests, in clearings, and sometimes in lawns, in spring, summer, and fall. *Amanita virosa* is widely distributed in North America; *A. verna* is rare on the East Coast, but occurs on the Pacific Coast, with records from Seattle, Washington and Vancouver Island, British Columbia.

TOXICITY The toxins of Death Angel are similar to those of Death Cap (*Amanita phalloides*) (p. 39), and the symptoms of poisoning are the same.

TREATMENT Same as Death Cap (*Amanita phalloides*) (p. 39.)

NOTES *Amanita verna* and *A. virosa* are classed by some authorities within a single species. Both are similar in habitat, and their poisonous qualities are the same. Other very similar species occurring in North America include *A. bisporigera,* widely distributed in the East, and *A. ocreata,* reported from California. These mushrooms are all strikingly beautiful, but deadly. In fact, mushrooms of the *Amanita phalloides—virosa—verna* group are the most poisonous species known. Their toxins, both amatoxins and phallotoxins, which are cyclic peptides, are more difficult to treat because they are delayed in their action. Symptoms do not usually appear for at least 6 hours after ingestion, and death occurs as long as 10 days later.

Sweat-causing Clitocybe, or Sweating Mushroom (Tricholoma Family)

Clitocybe dealbata (Tricholomataceae)

Fig. 13. Sweat-causing Clitocybe (*Clitocybe dealbata*).
AL FUNK

QUICK CHECK Small to medium-sized mushroom of open, grassy areas and lawns; cap dry, grayish white, smooth, slightly depressed at center, with irregular, incurved margins; gills slightly running down stem, which is short and tough. Very poisonous; sometimes fatal. Symptoms appear within 2 hours.

DESCRIPTION *Cap* grayish white when dry, to grayish tan when moist; 2–5 cm (0.8–2 in.) broad, at first convex, then flat and depressed at center at maturity; margin irregular, indented, and incurved or inrolled; *Flesh* thin, whitish, with mild taste; *Gills* white, close and narrow, attached or running down the stem slightly; *Stem* 1–7 cm (0.4–2.8 in.) long and up to 8 mm (0.3 in.) thick; solid, tough, same color as cap; sometimes curved, and often off-center; *Ring* not present; *Spore print* white.

OCCURRENCE Single to numerous on ground in open, grassy areas and sometimes woods; often on lawns, where it frequently forms rings. Found throughout North America from summer through fall.

TOXICITY This mushroom contains significant, and potentially fatal, amounts of the toxin muscarine (0.15 percent dry weight), the same compound present in White Inocybe (*Inocybe geophylla*), as well as other members of *Clitocybe* and *Inocybe*. For symptoms and treatment of muscarine poisoning, see under White Inocybe (*Inocybe geophylla*) (p. 52).

TREATMENT See under White Inocybe (*Inocybe geophylla*) (p. 52).

NOTES Sweat-causing Clitocybe is occasionally abundant on lawns, growing together with other lawn-inhabiting mushrooms such as Fairy Ring Mushroom (*Marasmius oreades*) and Meadow Mushroom (*Agaricus campestris*). Therefore, it is very important to be aware and able to recognize this very toxic mushroom. Other North American *Clitocybe* species known to contain muscarine and capable of causing poisoning include: *Clitocybe dilatata, C. morbifera,* and *C. rivulosa,* all described by Ammirati et al. (1985).

Deadly Conocybe, or Ringed Conocybe
(Bolbitius Family)

Conocybe filaris (syn. *Pholiotina filaris*)
(Bolbitiaceae)

Fig. 14. Deadly Conocybe (*Conocybe filaris*).
KIT SCATES BARNHART

QUICK CHECK Small, brown, gilled mushroom with long, thin stem; conspicuous ring halfway down stem. Deadly.

DESCRIPTION *Cap* smooth and tawny-brown, up to 2.5 cm (1 in.) across, cone-like to convex or flat, often with central knob; *Gills* broad and close together, notched, whitish to rust colored; *Stem* up to about 4 cm (1.5 in.) long, thin, and yellow-brown to orange-brown; *Ring* membranous and movable, central on stalk; *Spore print* cinnamon-brown.

OCCURRENCE Widely distributed in North America; scattered in lawns and grassy areas on rich, humus soils.

TOXICITY Recently it has been shown that this species contains the same lethal toxins, amanitins and phalloidins (cyclic peptides), as the deadly Destroying Angel (*Amanita verna—virosa* complex), and is very poisonous.

TREATMENT see under Destroying Angel (*Amanita verna—virosa* complex) (p. 43) and Death Cap (*A. phalloides*) (p. 39).

NOTES This mushroom is a good example of an "LBM" (little brown mushroom) that is highly toxic. People should refrain from eating all "LBMs" unless positive they are edible.

Inky-cap, Inky Coprinus, or Alcohol Inky
(Inky-cap Family)

Coprinus atramentarius
(Coprinaceae)

Fig. 15. Inky-Cap (*Coprinus atramentarius*).
IAN G. FORBES

QUICK CHECK Small to medium, grayish mushroom with rounded, cone-shaped to bell-shaped cap, changing into an inky black fluid as the mushroom matures. Often causes unpleasant toxic reaction if eaten with alcohol, but not fatal. Symptoms appear within 30 minutes after drinking alcoholic beverages and eating the mushroom.

DESCRIPTION *Cap* rounded, conical or bell-shaped, dull gray to grayish brown, smooth or mealy, ribbed radially, often splitting at the edges; up to 8 cm (3 in.) long ; *Flesh* thin, grayish white, with no distinctive odor; *Gills* crowded, free from stem, grayish white when young, becoming dark gray, then black and dissolving into an inky black fluid; *Stem* up to 12 cm (5 in.) long and 2 cm (0.8 in.) thick; covered with tiny, flattened hairs; hollow and splitting easily; *Ring* membrane-like, faint or sometimes conspicuous, located near the base of the stem; *Spore print* black.

OCCURRENCE On ground, in clusters; common in waste places, grassy areas, gardens, on road shoulders, and on debris. Common throughout North America from late summer to late fall.

TOXICITY Inky-cap contains varying amounts of an amino acid derivative, coprine, which interferes with the body's alcohol metabolism. The toxic effect of coprine takes place only in the presence of alcohol (ethanol); when no alcoholic beverages have been drunk, the mushroom is not toxic and is in fact considered edible.

A toxic reaction normally takes place within 30 minutes after ingesting Inky-cap and drinking alcohol, and lasts about 2 hours. A recurrence may be experienced if alcohol is taken again within the next few days. Major symptoms are: flushed face and neck, chest pains, palpitations, rapid heart beat, low blood pressure, feeling of swelling in hands and feet, profuse sweating, metallic taste in the mouth, nausea, vomiting, visual disturbance, weakness, dizziness, and rarely, difficult breathing or coma. Poisoning is not fatal, and recovery is usually fairly rapid and complete within a few hours. There is a considerable variation in the reaction of different individuals; some are totally unaffected by the toxin.

TREATMENT Empty stomach if ingestion of mushroom or alcohol was recent; follow with activated charcoal and saline cathartic; treat for shock; keep patient quiet and still; maintain fluid and electrolyte balance; if low blood pressure persists, treat with dopamine; monitor heart and blood pressure; maintain ventilation. The patient must avoid alcoholic beverages for at least a week after eating Inky-cap mushrooms.

NOTES Some other *Coprinus* species are known to contain coprine, but the well known edible species, Shaggy Mane (*C. comatus*), apparently does not. Like Inky-cap, it dissolves into a black fluid when old, but when young it is a choice edible species, with no toxic effects when eaten with or without alcoholic beverages.

A toxic reaction with alcohol, similar to that of Inky-cap, has been reported from Japan for an unrelated mushroom, *Clitocybe clavipes* (Fat-footed, or Club-footed Clitocybe), a species considered edible except when consumed with alcohol.

Coprine may become a useful drug in alcoholism therapy. Its action is similar to that induced by the drug disulfiram, currently used for treatment of alcoholics, but coprine has fewer side effects.

Fig. 16. Cortinarius (*Cortinarius* sp.). BILL McLENNAN

Cortinarius Species, or Corts
(Cortinarius Family)

Cortinarius spp.
(Cortinariaceae)

QUICK CHECK Fleshy mushrooms with rounded cap, young specimens with cobweb-like cortina veil extending from edge of cap to stem. Some species highly toxic; potentially fatal due to kidney failure. Symptoms may be delayed 2–17 days. For safety, do not eat any *Cortinarius* species.

DESCRIPTION There are hundreds of different species in this genus, with many variations in size, shape, and color, making *Cortinarius* species extremely difficult to identify. *Cap* fleshy, variable in color, rounded; a universal veil is often present, adhering to the surface with thin, silky fibrils, becoming slimy or glutinous when wet; spiderweb-like, partial veil (cortina), composed of delicate strands, extends from the edge of the cap to the stem, in young specimens usually covering the gills; veil disappearing at maturity; *Flesh* variable in color; *Gills* usually rusty brown in mature specimens; *Stem* variable in size, color; *Ring* veil often reduced to ring or series of partial rings on the lower part of the stem in older specimens; *Spore print* rusty brown to cinnamon brown. A detailed description of the genus and representative species may be found in Ammirati et al. (1985).

OCCURRENCE On ground; over 800 species occurring in North America. The thread-like mycelium forms a special type of symbiotic relationship (mycorrhizal) with the roots of trees, shrubs, and other plants. In fact, *Cortinarius* is the largest genus of mycorrhizal mushrooms in North America. *Cortinarius gentilis,* a toxic and common species in the genus, occurs abundantly under conifers and is widespread in North America.

TOXICITY Although some *Cortinarius* species are identifiable and a few are known to be edible, the facts that many are not well known or even described, and that some are highly toxic, make it advisable not to eat any of this group until more is known about them. Several orange-colored species are included in the highly toxic group, including *Cortinarius gentilis* (Deadly Cortinarius) and *C. orellanus* (Poznan Cortinarius). The latter is thus far known only from Europe. In 1950 a toxin was isolated from *Cortinarius orellanus* and named orellanin. It is composed of one or more cyclopeptides with bipyridyl structure. This toxin may also occur in *C. gentilis*, but this has not yet been proven.

   Cortinarius poisoning is potentially fatal, due to kidney failure. The liver and nervous system may also be affected. Symptoms may be delayed from 2–17 days after ingesting the mushrooms. Nausea and vomiting, followed by sweating, shivering and stiffness, pain in the limbs and abdomen, constipation or diarrhea, severe thirst, reduction and then increase in urine production, sleepiness, and convulsions are all known symptoms.

TREATMENT For initial consumption: induce vomiting, or perform gastric lavage; follow with activated charcoal. For severe or advanced cases, treat for symptoms; corticosteroids, peritoneal dialysis, and hemodialysis to prevent kidney failure; high protein diet, and intravenous doses of protein hydrolysate help prevent liver damage; kidney transplantation may be required to avoid fatality.

NOTES Several fatalities caused by eating *Cortinarius* are reported from Great Britain and Europe, but so far no fatal cases are known in North America. Because of the long delay in appearance of symptoms, it is quite possible that poisoning has occurred in North America but simply has not been diagnosed. At least one species of *Cortinarius* has caused fatal poisoning in sheep; all should be considered toxic to animals as well as humans.

   Cortinarius toxins are all the more insidious because of the long period without symptoms. Unless the physician is aware that this type of mushroom is involved, when the symptoms do appear an incorrect diagnosis may be made and inappropriate treatment given.

Autumn Galerina
(Cortinarius Family)

Galerina autumnalis (syn.
  Pholiota autumnalis)
(Cortinariaceae)

Fig. 17. Autumn Galerina (*Galerina autumnalis*).
AL FUNK

QUICK CHECK Small to medium-sized mushroom on decayed wood in forests (or lawns if growing over buried wood); smooth, brownish, broadly bell-shaped cap; hairy, band-like ring around stem. Highly toxic, potentially fatal. First symptoms appear in 6–24 hours, followed by period of apparent improvement, then by symptoms of liver and kidney failure after 4 or more days.

DESCRIPTION *Cap* 2–6 cm (0.8–2.5 in.) across, bell-shaped, becoming rounded with a small, central knob; surface smooth, sticky when wet, light tan (when dry) to dark brown (when wet), with distinct lines, or striations, on the margin when wet; *Flesh* light brown, watery, without distinctive odor; *Gills* rusty, attached to stem; *Stem* up to 7 cm (3 in.) long and 6 mm (0.2 in.) thick, slightly enlarged at the base and tapering upward; brown, covered with whitish hairs; *Ring* band-like, narrow, white-hairy, sometimes inconspicuous; *Spore print* rusty brown.

OCCURRENCE Common in woods from spring to late fall, sometimes early winter; scattered or in dense clusters, growing on well-decayed wood, both conifer and hardwood, or sometimes on buried wood (even in lawns), chips, or sawdust. Particularly abundant after a heavy rain.

TOXICITY This mushroom contains the same groups of toxins, the fatally poisonous phallotoxins and amatoxins, as Death Cap (*Amanita phalloides*) (p. 39). The first symptoms appear after 6–12 hours or even longer. Nausea, vomiting, diarrhea, and severe abdominal cramps are followed by a period of improvement lasting up to 4 days. Then liver and kidney failure ensue, frequently leading to coma and death in 7–10 days.

TREATMENT Follow the same procedures as for Death Cap (*Amanita phalloides*) (p. 39). Treatment should start immediately. Correct identification of the mushroom is vital.

NOTES The genus *Galerina* includes about 200 species of small, hard to identify mushrooms, some deadly poisonous, others with unknown edibility. All should be strictly avoided. Two closely related North American species, *G. venenata* and *G. marginata*, are similar in appearance and habitat to Autumn Galerina and just as deadly. The former is known only from the Pacific Northwest and grows on lawns; the latter is widely distributed in forested areas.

Parents of small children and pets should watch for the Galerinas on lawns, especially if there is buried wood beneath. Although they would scarcely be sought by mushroom pickers, children might notice them and eat or chew on them. Furthermore, both of these species can easily be confused by the inexperienced collector for the hallucinogenic "magic" mushrooms, *Psilocybe* and its relatives. People who collect the latter should make themselves completely familiar with toxic look-alikes.

False Morel, Brain Mushroom, or Turban Fungus
(Cup Fungi)

Gyromitra esculenta
(Helvellaceae)

Fig. 18. False Morel (*Gyromitra esculenta*).
MARY W. FERGUSON

QUICK CHECK Brownish mushroom of woods and gardens with pale, whitish stem and irregularly folded, convoluted cap; very variable in shape and color. Some strains are highly toxic, especially if eaten raw; potentially fatal due to liver damage. Symptoms appear in 6–12 hours. Do not confuse with true, or Edible, Morel.

DESCRIPTION *Cap* tan to reddish brown or chocolate-brown; up to 10 cm (4 in.) high and 15 cm (6 in.) wide; irregular in shape, contorted, folded, and wrinkled, with brain-like ridges (but not pitted as in true morels); *Flesh* brittle, pale; *Stem* up to 6 cm (2.5 in.) high, smooth, fragile, thick and solid at the base, becoming somewhat grooved, hollow, or with open chambers above; pale flesh-colored; *Ring* none; *Spore print* yellowish. *Gills* are not present in ascomycetes.

OCCURRENCE Grows from early spring to early summer, singly or in small groups on ground in open, moist, wooded areas under coniferous trees, and in gardens and compost heaps. Widely distributed throughout North America and in the temperate zone of the Northern Hemisphere.

TOXICITY False Morel is potentially fatal, especially if eaten raw. The main toxic component is called gyromytrin. This compound is very unstable and is easily converted at moderate cooking temperatures to toxic monomethyl hydrazine, itself very unstable and volatile. Another substance, called helvellic acid, isolated in the 1930s, was probably a crude product containing gyromytrin. The lethal dose in humans has been estimated at between 10 and 50 mg/kg. Symptoms usually appear 6–12 hours after eating, but cases with symptoms occurring after only two hours are known. Symptoms may include bloated feeling, nausea, vomiting, severe diarrhea, abdominal cramps, headache, dizziness, fever, difficulty breathing, possible rapid heart beat and low blood sugar, and coma. Death, due to kidney and liver damage, may occur within a few days. The toxin accumulates in the body and until a certain threshold is reached few or no symptoms appear. Consumption of the mushroom once may not induce symptoms, but a second or third meal may produce a severe poisoning. Since the toxin is volatile, it may be removed by parboiling but this is not guaranteed. False Morel is often eaten after such treatment. We strongly recommend it not be eaten under any circumstances.

TREATMENT Induce vomiting, or perform gastric lavage; follow with activated charcoal and a saline cathartic; treat for symptoms: maintain fluid and electrolyte balance. The use of pyridoxine hydrochloride (25 mg/kg over 3 hours) has been suggested as an antidote for gyromytrin poisoning; treat seizures with i.v. diazepam; oxygen and i.v. methylene blue may counteract methemoglobinemia.

NOTES This mushroom is named from its similarity to the true, or Edible, Morel (*Morchella* spp.), a well known edible mushroom of springtime, whose stem is hollow, not chambered, and whose dark grayish, elongated cap is pitted with vertically aligned, honeycomb-like cavities.

The toxicity of False Morel in North America has caused some confusion, and it appears that the mushroom is variable in its poison content and also causes different reactions in different individuals. For example, it is considered edible in western North America, where it is often eaten without toxic effects. In the Midwest and eastern North America, it has caused several poisonings, with at least seven fatalities known. It has been proposed that different chemical strains of this species exist, a theory that would explain its variable toxicity. It has been suggested also that specimens collected at higher elevations in the West have lost the toxin due to its volatility and the lowered atmospheric pressure.

False Morel is said to be safe to eat if dried or boiled for 10 minutes and the water discarded. Adam Szczawinski has eaten young, firm specimens of this species in both eastern Europe and western North America without any toxic reaction. Until more is known about this mushroom, however, the recommendation is to leave it strictly alone.

White Inocybe, or White Fiber Head and Its Relatives
(Cortinarius Family)

Inocybe geophylla and related species
(Cortinariaceae)

Fig. 19. White Inocybe (*Inocybe geophylla*).
R. & N. TURNER

QUICK CHECK Small, silvery white to light tan or lilac mushrooms of coniferous woods and nearby clearings; small knob at center of cap. Very poisonous; sometimes fatal. Symptoms appear within 2 hours.

DESCRIPTION *Cap* silvery white or occasionally light tan or lilac, dry, with a silvery sheen; 1.5–3.5 cm (0.6–1.4 in.) broad, conical to bell-shaped when young, expanding to nearly flat, with a prominent central knob, or umbo; splitting at the margin at maturity; usually a web-like, partial veil, or cortina, extends from the edge of the cap to the stem; *Flesh* thin, white, of no distinctive odor or taste; *Gills* close, joined to the stem, white when young, becoming gray, then brown; *Stem* up to 6 cm (2.4 in.) long and 5 mm (0.2 in.) thick; cylindrical, covered at the top with tiny white hairs; *Ring* hairy, band-like, often inconspicuous; *Spore print* rusty-brown.

OCCURRENCE On damp soil, singly or in small groups, in coniferous or mixed forest; sometimes on open ground or lawns under coniferous trees. Found throughout North America and Europe from fall to early winter.

TOXICITY This mushroom contains the toxin muscarine, a quaternary ammonium compound, which may cause serious illness or death. It inhibits the conduction of impulses between nerve cells. Symptoms, which occur from 15 minutes to 2 hours after ingestion, include copious salivation, profuse sweating, tears, nausea, chest spasms, abdominal pain, and slowed pulse. Vomiting, watery diarrhea, reduced blood pressure, slow heart beat, asthmatic wheezing, and blurred vision may also occur. A mortality rate of about 5% has been estimated.

TREATMENT Empty stomach; follow with activated charcoal and a saline cathartic. Atropine is a specific antidote for muscarine poisoning, but use only if symptoms are pronounced. Sufficient quantities should be given to produce dryness of the mouth. Large quantities, producing capillary dilation, which have been previously advocated, are usually not necessary and may cause symptoms of atropine poisoning (delirium and hyperthermia) in small children (recommended dose: 0.5 mg–1.0 mg i.v. in adult; 0.05 mg/kg i.v. in child); apply general supportive measures; maintain fluid and electrolyte balance, and assist breathing if necessary.

NOTES This is one of the most common *Inocybe* species in North America, as well as in Europe. The genus includes about 300 small or medium-sized mushrooms, many very similar in appearance. They are characterized by a convex cap with a central knob. The upper surface of the cap is often covered with radiating fibrils or scales, these brown in many species. *Gills* are clay-colored, and spores yellowish to light brown. Identification to species is difficult, and may require microscopic technique. Many species of this genus contain muscarine and can cause serious illness. Besides *I. geophylla,* another widely distributed toxic North American species is *I. rimosa* (syn. *I. fastigiata*); several others are described by Ammirati et al. (1985). The lilac-colored form of *I. geophylla* is treated by some mushroom experts as a separate species, *I. lilacina.* Mushroom collectors should avoid the entire group.

Muscarine was first isolated from Fly Agaric (*Amanita muscaria*), which contains insignificant amounts. It is found in much higher concentrations in species of *Inocybe* and *Clitocybe.*

Jack O'Lantern, Copper Trumpet, or False Chanterelle
(Tricholoma Family)

Omphalotus illudens (often cited as *O. olearius*) and related species
(Tricholomataceae)

Fig. 20. Jack O'Lantern (*Omphalotus illudens*).
KIT SCATES BARNHART

QUICK CHECK Medium-sized, orange to yellowish orange, luminescent mushroom on decaying wood, with dry cap, and sharp-edged gills which extend down the stem and glow in the dark. Very poisonous; occasionally fatal. Symptoms appear within two hours. Do not confuse with edible Chanterelle.

DESCRIPTION *Cap* 5–20 cm (2–8 in.) across, rounded to flat and usually depressed in the middle, with a shallow, central knob; deep orange to yellowish, fading with age; upper surface smooth and dry (not sticky); margin enrolled at first, then upturned and wavy; *Flesh* whitish and firm, with no distinctive odor; *Gills* yellow-orange, sharp-edged, luminescent (glowing in the dark), running down the stem; *Stem* light orange, 7–25 cm (3–10 in.) high and 5–22 mm (0.2-0.9 in.) thick, tapering at the base; *Ring* not present; *Spore print* white to pale cream or pale yellow.

OCCURRENCE Growing in clusters on decaying wood, including standing trunks, fallen logs, or buried wood, sometimes fruiting in lawns by trees with heart rot; most commonly associated with oak (*Quercus*). Occurs in Canada from the Great Lakes region eastward, but is not common. Found throughout the eastern United States, and into the South. Two other species, also toxic, occur in North America: *Omphalotus olivascens*, reported from California, and *O. subilludens* in the southeastern United States.

TOXICITY The toxins of this mushroom and its relatives, though potentially deadly, have not yet been fully identified. It has been reported to contain muscarine-like toxins, akin to those of some *Clitocybe* and *Inocybe* species, but recently other toxins have been identified—the sesquiterpenoids illudin-s (lampteral) and illudin-m. Symptoms of poisoning occur from 15 minutes to 2 hours after ingestion: salivation, sweating, tears, nausea, chest spasms, abdominal pain, slowed pulse, reduced blood pressure, vomiting, watery diarrhea, asthmatic wheezing, and blurred vision all may occur in severe cases. Additionally, tingling fingertips and an unpleasant, metallic taste in the mouth have been noted.

TREATMENT Treat for symptoms; in cases of vomiting and diarrhea, maintain fluid intake and electrolyte balance.

NOTES Jack O'Lantern is sometimes confused with the edible Chanterelle mushroom (*Cantharellus cibarius*), which is close to it in color, but can be easily distinguished by the gills, which are shallow ridges and branched in Chanterelle while sharp-edged and not branched in the Jack O'Lantern mushroom. Furthermore, unlike Jack O'Lantern, Chanterelle does not grow on wood. The eerie, greenish luminescence of Jack O'Lantern's gills in the dark often lasts up to two days after it is collected but old specimens may not luminesce. *Omphalotus illudens* is commonly, but mistakenly, called *O. olearius*, and some include it in the genus *Clitocybe*, as *C. illudens*. In Japan a close relative of *Omphalotus* spp., *Lampteromyces japonicus*, sometimes causes deaths.

**Common Large Psilocybe, or
Magic Mushroom**
(Stropharia Family)

Psilocybe cubensis
(Strophariaceae)

Fig. 21. Common Large Psilocybe (*Psilocybe cubensis*), grown in mason jar. KIT SCATES BARNHART

QUICK CHECK Medium-sized mushroom of well-manured fields, with bell-shaped cap which is chestnut-brown when young, lighter at maturity; ring white, membranous and persisting, and stem staining blueish when bruised. Hallucinogenic, causing confusion, delirium, and visual disturbances within a hour, with recovery in 6–18 hours; no fatalities known, but do not confuse with highly toxic Autumn Galerina.

DESCRIPTION *Cap* 2–8 cm (0.8–3 in.) across, conical to bell-shaped, with small knob, or umbo, at tip; expanding to flat-topped with age; chestnut-brown, becoming light brown at maturity; staining blueish when bruised or injured; smooth and sticky when wet; *Flesh* first, white, bruises blue; *Gills* narrow, close, gray, becoming purplish gray to almost black, with distinct, white edges; attached to the stem, and notched at point of attachment; *Stem* 5–15 cm (2–6 in.) long, 5–15 mm (0.2-0.6 in.) thick, smooth, whitish, enlarged at the base, and bruising blueish green; *Ring* white, membranous, persistent; *Spore print* purplish brown.

OCCURRENCE Scattered or in groups in grassy meadows, mostly on cow or horse manure, or on well manured ground. Found throughout the southern United States and Mexico, from spring to late fall.

TOXICITY This species contains the same hallucinogenic compounds, psilocin and psilocibin, found in the following species, *Psilocybe semilanceata*. Symptoms of intoxication are the same.

TREATMENT Same as *Psilocybe semilanceata* (p. 56).

NOTES This mushroom is widely used as an hallucinogen, and is said to be the most widely cultivated mushroom in North America. As with the previous species, the main danger of its use is in misidentification, because it closely resembles other, much more toxic species.

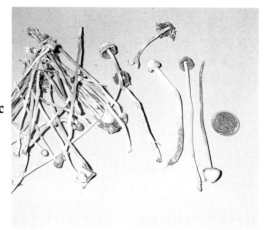

Liberty Cap, or Magic Mushroom
(Stropharia Family)

Psilocybe semilanceata
(Strophariaceae)

Fig. 22. Liberty Cap (*Psilocybe semilanceata*). AL FUNK

QUICK CHECK Small, conical-capped, brownish mushroom of wet grasslands, especially on cow dung; cap with distinct knob at tip; a brown, cobweb-like ring around the stem is evident in young specimens, but quickly deteriorates with age. Hallucinogenic, causing confusion, delirium, and visual disturbances within an hour, with recovery in 6–18 hours; no fatalities known, but do not confuse with highly toxic Autumn Galerina.

DESCRIPTION *Cap* up to 2.5 cm (1 in.) across, sharply conical with prominent knob, or umbo, at tip; semitransparent, usually dark chestnut-brown when moist, drying to light tan or yellowish, occasionally olive-tinted or with a blueish cast; curved under at edges when young; surface sticky and gelatinous when wet; *Flesh* thin, watery, pale tan to brown; *Gills* narrow, close, attached to stem, and not notched at point of attachment; pale tan at first, becoming brown, then purplish brown with pale edges; *Stem* up to 10 cm (4 in.) long and 25 mm (0.1 in.) thick, smooth, flexible, light brown, tending to dark brown at base; injured or bruised spots may turn blue; *Ring* delicate, brown, cobweb-like, quickly disappearing, leaving a sticky zone darkened by the color of the spores; *Spore print* purplish brown.

OCCURRENCE Scattered or in groups on wet meadows and grasslands, often on cow dung. Common throughout the Pacific Northwest, west of the Cascade Mountains, from British Columbia to California, also known from the eastern Maritime Provinces in Canada. Found in fall and early winter, occasionally in spring in Oregon and Washington. It also occurs commonly in Europe and in the United Kingdom, where it is known as Liberty Cap.

TOXICITY This mushroom contains the hallucinogenic toxins, psilocin and psilocybin, which are indole derivatives. Their effects on people are similar to those of LSD (lysergic acid diethylamide). Hallucination can occur from eating the mushrooms raw or cooked, or from drinking broth made by boiling them.

Symptoms of intoxication, becoming evident after 30–60 minutes, are: confusion, tremors, visual disturbance, palpitations, respiratory difficulties, and feelings of anxiety, paranoia, or euphoria. Varying degrees of delirium, laughter, and visual aberration of speed, light, and color may also occur. Recovery normally takes place in 6–18 hours, depending on the individual, his condition, and the quantity of mushrooms ingested. The mushroom is not known to be fatal, but can be mistaken for other, more deadly types.

TREATMENT The toxins are usually eliminated from the body within 4–6 hours, and specific therapy is seldom required. If the patient is panicky, a tranquilizer (such as diazepam or chlorpromazine) can be given. Otherwise, only general supportive measures are needed.

NOTES Hallucinogenic mushrooms containing psilocin and psilocybin were used some 2000 years ago by the Aztecs of Mexico in their religious rituals, and were considered "flesh of the Gods." Within the last two decades or so, these mushrooms have become popular among some groups of people, particularly teenagers and young adults, as "recreational" drugs. The species most frequently used are *Psilocybe semilanceata, P. pelliculosa,* and *P. cubensis.* Altogether, there are over 70 hallucinogenic species in the genus, with some 20 or more occurring in North America.

Species in a number of other mushroom genera, including *Conocybe, Paneolus,* and *Stropharia,* also contain psilocin and psilocybin, but some contain more deadly toxins as well. A major problem for potential users of the drugs is the danger of misidentification. From time to time, highly toxic mushrooms such as *Galerina autumnalis,* or species of *Paneolus* or *Stropharia,* have been eaten accidentally, sometimes with fatal results.

Emetic Russula and Its Relatives
(Russula Family)

Russula emetica and related species
(Russulaceae)

Fig. 23. Emetic Russula (*Russula emetica*). IAN G. FORBES

QUICK CHECK Showy, medium-sized mushroom of mossy woods; bright red cap, whitish, brittle stem and gills, and acrid, peppery taste.

Causes vomiting and digestive distress within an hour of eating, but is not fatal.

DESCRIPTION *Cap* up to 10 cm (4 in.) across; rounded at first, becoming flattened, with depressed center and upturned margins; usually bright red, sometimes mottled yellowish and red; smooth, and sticky when wet; margin marked by radiating lines, or striations, which fade with age; *Flesh* white to off-white, brittle, and very acrid and peppery to the taste; *Gills* free, or almost free, from the stem; white to yellowish white, and thin; *Stem* up to 10 cm (4 in.) high and 2.5 cm (1 in.) thick, enlarging toward the base; whitish; smooth and dry, becoming hollow with age; *Ring* lacking; *Spore print* white to yellowish white. There are several related Russulas, including *R. rosacea, R. sanguinea,* and *R. fragilis,* which are also acrid and should be avoided. Another relative, *R. vesicatoria,* causes blistering of the lips and tongue when tasted.

OCCURRENCE Grows from late summer to fall singly or in groups on the ground in mossy areas of coniferous or mixed woods; rarely on well-decayed wood or in peat bogs. Common and widely distributed in North America.

TOXICITY The toxins are as yet not well known, and not all people are affected to the same degree. Symptoms, which usually occur within 30–45 minutes after eating the mushroom, are: vomiting, nausea, abdominal cramps, and diarrhea. The mushroom is named after the inevitable vomiting and diarrhea it causes. In severe cases, discomfort persists for 24 hours or more, but normally recovery is within a few hours.

TREATMENT General supportive treatment for symptoms.

NOTES Russulas are among the most abundant and conspicuous of North American mushrooms, but are difficult to identify to species. According to some, even Emetic Russula and its relatives become edible and palatable after thorough cooking, but until more is known about their toxins all Russulas with an acrid taste should be avoided.

Poison Puffballs
(Fall Puffball Family)

Scleroderma spp.
(Sclerodermataceae)

Fig. 24. Pigskin Poison Puffball (*Scleroderma aurantium*). MARY W. FERGUSON

QUICK CHECK Thick-skinned, smooth to warty, light-brown puffball-like mushrooms, with purplish black spore mass inside at maturity. Cause nausea and stomach upset, but not fatal. Do not confuse with edible Puffballs.

DESCRIPTION Spherical to flattened puffball-like mushrooms, rooted at the base, without separate cap or stem; up to 10 cm (4 in.) across and 4 cm (1.6 in.) high, sometimes resembling small, rounded potatoes; covered with smooth to rough and warty, rind-like skin, which is yellowish tan to dark brown; the outer wall breaks irregularly at the top at maturity to expose the spores; *Flesh* of skin white; of spore mass, at first white, but soon becoming purplish black to black. A number of species occur in North America, and they are often difficult to distinguish.

OCCURRENCE Poison Puffballs are widely distributed in North America, growing from summer through fall on the ground on wood debris, in woods, and sometimes gardens, usually near trees.

TOXICITY *Scleroderma* species are known to cause nausea, vomiting, abdominal cramps, and diarrhea, but their toxins have not been identified. Symptoms appear about 30 minutes after eating. The degree of reaction depends on the individual, but can be severe. Recovery is usually fairly rapid.

TREATMENT Induce vomiting; treat for symptoms.

NOTES Poison Puffballs can be confused with true Puffballs (*Lycoperdon* spp., and other, related genera), which are favourite edibles. To distinguish edible Puffballs, slice them in half from top to bottom; a true Puffball at the edible stage will look and feel like a marshmallow, firm but soft, white, and homogeneous (without gills or discrete stem). Puffballs too old for eating will be yellowish or greenish brown and very soft inside, becoming powdery when dry. Poison Puffballs at maturity will be firm and purplish black inside, with a thick, rind-like skin. Although some species of *Scleroderma* have been eaten as truffle substitutes, their taste is rather acrid, and all should be avoided because of their potential harmful properties.

References for Chapter II: Ainsworth 1952; Allegro 1970; Ammirati et al. 1985; Anon. 1989; Bandoni and Szczawinski 1976; Brick 1961; Canadian Pharmaceutical Association 1984; Cooper and Johnson 1984; Emboden 1979; Faulstich 1980; Faulstich et al. 1980; Floersheim 1987; Groves 1979; Haard and Haard 1980; Hardin and Arena 1974; Hatfield 1979; Hatfield and Brady 1975; Hotson and Lewis 1934; Kingsbury 1964; Lincoff 1981; McKenny 1971; Miller 1980; Stamets 1978; Wasson 1968.

CHAPTER III

POISONOUS PLANTS OF WILD AREAS

This book cannot replace the advice and assistance
of qualified medical personnel.
In all cases of suspected poisoning by plants, or any other substance,
immediate qualified medical advice and assistance should be sought.

ALGAE

QUICK CHECK   Shellfish poisoning from microscopic algae can be fatal. Almost all of the larger seaweeds are harmless. Algae and bacteria can contaminate fresh water; do not drink any water not known to be safe.

Never harvest shellfish for eating without checking with those who know local water conditions, such as fisheries or health authorities. If shellfish poisoning is suspected, contact physician immediately. Keep samples of water and shellfish for identification.

What Are Algae?

Algae are members of several large groups of simple, mainly aquatic plants. They include the so-called seaweeds, which are large, easily visible marine or freshwater algae, as well as many microscopic types that in large numbers produce greenish or brownish scum or slime in ponds and tidepools. Many algae, particularly seaweeds, are edible and are an important part of the human diet in many areas of the world, including Japan, China, Hawaii and the South Pacific. Seaweeds are also eaten by Native peoples of coastal North America, and some species, such as Dulse (*Palmaria palmata;* syn. *Rhodymenia palmata*), are widely marketed in Canada and the United States.

Poisonous Freshwater Algae

Some algae can be toxic to humans and animals. Among fresh-water algae, only a few belonging to the bluegreen algae (Division

Cyanophyta, Class Myxophyceae), are known to cause poisoning. Bluegreen algae are common in almost any body of water. Although they are mostly microscopic, under some conditions they can grow rapidly and accumulate in large masses in the upper layers of water, forming a "bloom" or scum. Only a few species of algae that form these "blooms" are poisonous, and since several or many types often grow together, the actual toxic species are difficult to isolate. Furthermore, toxicity varies greatly over brief time spans as short as a few hours. Slight changes in environmental conditions may dissipate concentrations of the harmful algae quite suddenly. Several species have been implicated, but of these, only three are cited by Kingsbury (1964) as being well substantiated toxic organisms: *Anabaena flos-aquae*, *Aphanizomenon flos-aquae*, and *Microcystis aeruginosa*. Surprisingly, the toxin from *Anabaena flos-aquae* has been identified as the same as that from "red tide." These three types can be found in quiet, warm, nutrient-rich waters in North America and throughout the world.

Extensive loss of life and severe sickness of humans and many kinds of animals have been associated with drinking water contaminated with algal blooms in the northern United States and Texas and in southern Canada. One moderately poisonous algal toxin that has been isolated is a cyclic polypeptide containing seven amino acids. Others have not been identified, and may be products of decomposition or of bacteria associated with the algae. Symptoms of algal poisoning appear rapidly, within 15–45 minutes of drinking contaminated water, and include nausea, vomiting, abdominal pain, diarrhea, prostration, muscular tremors, difficulty in breathing, discoloration of the skin due to lack of oxygen, general paralysis, convulsions, and, occasionally, death within 1–24 hours. Bacteria and other organisms in fresh water may also cause illness. The best policy is not to drink stagnant water, or any water not known to be safe.

Shellfish Poisoning: "Red Tide"

For marine algae, the most serious cause of poisoning is through eating filter-feeding molluscs such as mussels, clams and oysters, and sometimes their predators that have accumulated the toxin from a large quantity of a tiny microscopic alga of the dinoflagellate group (mainly of the genus *Protogonyaulax*, Division Pyrrophyta, Class Dinophyceae). Under conditions difficult to predict, populations of these minute plants multiply to concentrations of several million cells per liter of seawater. In these quantities, they may color the water brownish or reddish, giving rise to the term "Red Tide." It should be noted, however, that there are many organisms that cause a reddish coloration in the water that are not poisonous, and *Protogonyaulax* algae in concentrations enough to be toxic do not always produce a visible coloration.

"Red tide" *Protogonyaulax* species have over recorded history caused hundreds of cases of severe poisoning and death on the west coast of North America, from Alaska to California, and to a lesser degree on the east coast as well. Two species, *P. catenella* on the West Coast and *P. tamarensis* on the East Coast, are the main culprits. Saxitoxin is the principal toxin in about a dozen structurally related paralytic shellfish toxins based on a complex perhydropurine skeleton. These water soluble, extremely stable toxins act on the neuro-muscular system in a manner similar to that of curare, the South American arrow poison. Eating only a very few contaminated clams or other bivalves may be lethal for a person. Numbness, nausea, difficulty in walking, thick speech, headache and increasing general paralysis are typical symptoms. Death is from respiratory failure. If the patient survives the first 24 hours, however, his prognosis is good.

In the winter of 1987 a new form of shellfish poisoning originating on the East Coast was attributed to accumulation in the water of a diatom, *Nitzschia pungens,* and the concentration of its deadly toxin, domoic acid, in mussels. Over 150 people became sick after eating mussels from the Canadian Maritimes region, and there were at least three deaths.

The best prevention for shellfish poisoning is to inquire about local tidewater conditions before harvesting shellfish. The summer months are particularly suspect. Removal of the viscera reduces, but does not eliminate, the level of toxicity. Many clams concentrate the toxin in the pigments at their siphon tips, possibly as a defense against predators which often nip them off. Commercially marketed shellfish are vigorously and effectively inspected through governmental agencies. The main victims of shellfish poisoning are uninformed individuals harvesting their own shellfish. Further, detailed information on shellfish and other types of seafood poisoning is provided in an article by Schantz (1973) and in the book by Tu (1988).

Seaweeds: Most Are Harmless

Of the larger, more easily visible marine algae, or seaweeds, only a few are toxic if eaten in quantity, and members of only one genus, *Lyngbya,* are potentially lethal. *Lyngbya* is a hair-like, filamentous bluegreen alga forming dense irregular floating mats or growths around atolls and in salt marshes, as well as in fresh water. Many types of fish feeding on *Lyngbya* or on other fish which feed on it become poisonous to humans. This type of poisoning, known as ciguatura poisoning, is prevalent in tropical regions such as the Caribbean and the South Pacific, and studies aimed at detecting and controlling it are ongoing.

Another seaweed which could cause problems to those trying to eat it is *Desmarestia*, a brown alga (Division Phaeophyta, Class Phaeophyceae) which has a very high acidity and contains esters of sulphuric acid. Members of this genus can cause severe digestive upset, but since they taste very sour they are unlikely to be eaten in any quantity.

References for Algae; Chapter III: Kingsbury 1964; Madlener 1977; Schantz 1973; Szczawinski and Turner 1980; Tu 1988.

Fig. 25. *Desmarestia herbacea,* a seaweed which produces sulfuric acid, rendering it inedible. R. & N. TURNER

FUNGI

QUICK CHECK Ergot, a small, hard, black elongated structure that grows on and contaminates grains, is highly toxic to humans and animals, when eaten in quantity or in small doses over a period of time; rarely a problem for humans in North America today, but animal poisoning occasionally severe. Some molds are poisonous and potentially carcinogenic.

If home-grown grains are eaten, check for ergot contamination. If poisoning from ergot-contaminated feed is suspected (see symptoms, below), have a sample checked by local agricultural authorities. Never eat rotting or moldy food, or feed moldy fodder to animals. Do not breath in dust from moldy hay or foods. Avoid old or improperly stored grains and nuts.

Ergot

In addition to poisonous mushrooms, which are treated in a separate chapter (Chapter III), there are many other types of fungi that cause poisoning to humans and animals. One of the most notorious is Ergot (*Claviceps* spp.), including several species of ascomycete fungi with complex life cycles that parasitize the grains of various cultivated and wild grasses. Most prominent is *C. purpurea,* which grows on rye, wheat, and barley. The resting phase of ergot is as a hard, blackish, elongated structure called a sclerotium. It is occasionally found as a contaminant in grain, causing crop loss by reducing the yield and quality of grain. (Grain containing more than 0.3 % ergot by weight is prohibited from sale due to potential toxicity.)

Ergot poisoning was known for many centuries under the name "St. Anthony's Fire," after the saint who is said to have suffered from it. Russian peasants often developed a chronic form of ergot poisoning, and people who ate bread made from infected grain suffered from a disease known as ergotism. There are still occasional outbreaks of the disease in some parts of the world, but in North

Fig. 26. Ergot (*Claviceps purpurea*) growing on Rye.
R. & N. TURNER

America human poisoning has virtually ceased. Animal poisoning, especially cattle, however, has occasionally been severe (Kingsbury, 1964).

Over 40 alkaloids, many known to be toxic, have been isolated from ergot. All are related chemically, being derivatives of lysergic acid. They are present in varying amounts depending on the strain and geographic location of the ergot sclerotia. Some are useful medically in stimulating uterine contractions during labor and in controlling uterine hemorrhage, but used incorrectly they can cause hallucinations and insanity.

Symptoms of ergot poisoning include irritation of the digestive tract (abdominal pain, with nausea, vomiting, diarrhea, and thirst), headache, loss of balance, lack of coordination, muscle tremors, and convulsions. These symptoms are followed by drowsiness and temporary paralysis. Large amounts of ergot ingested per day produce convulsive ergotism: hyperexcitability, belligerency, trembling, and convulsions. Ingesting small amounts of ergot over a period of weeks may cause chronic poisoning characterized by dry gangrene of the extremities. Irregular heart beat, variable blood pressure, and kidney failure sometimes occur. Humans and animals are affected similarly.

Treatment for ergot poisoning should include symptomatic and supportive measures, and, where necessary, vasodilators such as sodium nitroprusside. Amyl nitrite inhalations may be required. Convulsions may be treated with i.v. diazepam. For affected animals, care should be taken to avoid undue excitment in moving them away from infected vegetation.

Mold Poisons, or Mycotoxins

Another type of poisoning results from consuming certain compounds, called mycotoxins, from some microscopic fungi commonly known as molds. Molds sometimes contaminate food and can cause poisoning in people and animals. Molds have been recognized for centuries, and some have been used in the production of cheeses such as Roquefort and Camembert, and of antibiotics such as penicillin. (Even Camembert and other cheeses can be harmful when eaten together with certain tranquilizers and antidepressants.) However, it has been shown recently that many molds produce poisons.

A major impact of mold poisons arises from contaminated animal forage. So-called "forage poisoning" has caused illness and death of tens of thousands of cattle, horses, swine, poultry, and other domesticated animals in North America. Several toxic mold species, especially in the genera *Penicillium*, *Aspergillus*, and *Monascus*, have been incriminated through poisoning investigations and experimental feeding. The main losses have been associated with moldy corn.

Diseases caused by mycotoxins are known as mycotoxicoses. Originally, many were blamed on the feed plants themselves rather than on their mold contaminants. Mycotoxins vary considerably in chemical structure and belong to several different chemical groups. In 1961 researchers isolated a common mold, *Aspergillus flavus*, from peanut meal that had been responsible for the death of thousands of turkeys in Britain. Subsequently, it was found that this species and another, *A. parasiticus*, produced a mycotoxin known as aflatoxin, which caused tumors of the liver. Since that time, investigations on mycotoxins and mycotoxicoses have been numerous, and are ongoing.

The tumor-causing properties and other toxic effects of aflatoxin and some other mold poisons certainly have implications for human health. Differences in geographical distribution of liver cancer in humans have been attributed to different levels of occurrence of mycotoxins in food in various parts of the world, although this has not actually been proven. Improved harvesting and storage procedures for crops such as peanuts and grains have significantly reduced the risks of mycotoxicosis in humans.

It should be mentioned that food may contain mycotoxins even when not visibly moldy. Furthermore, the presence of mold on stored food does not necessarily indicate the presence of mycotoxins, which are produced only by some mold species. Today, many mycotoxins have been identified from mold species on a wide variety of foods such as cereal and nuts, and also from the milk and meat of animals that have eaten contaminated feed. There is also potential danger from inhaling the spores of many molds, which can produce an allergic respiratory condition in man and animals. A number of species in the genus *Fusarium* are known to cause this problem.

References for Fungi; Chapter III: Literature reviews given by Ciegler 1975; Patterson 1982; Wyllie and Morehouse 1978. See also Canadian Pharmaceutical Association 1984; Cooper and Johnson 1984; Kingsbury 1964; Miller 1973.

LICHENS

QUICK CHECK Lichens—small, moss-like or flattened plants of various colors growing on rock, bark, or wood—are mostly indigestible and can be acrid and irritating to the digestive tract, but few are seriously poisonous in small quantities. Some lichens can cause allergic skin reactions in some people.

What Are Lichens?

Lichens are complex structures consisting of two separate organisms, an alga and a fungus, growing together in a close, mutually beneficial relationship. They have several different growth forms. Some are crustose, adhering tightly to rock or bark. Others are thallose, with flattened, almost leaflike surfaces. Still others are fruticose, with stiff upright or hanging, branching structures. Some lichens are edible (including Black Tree Lichen, *Bryoria fremontii* and Rock Tripe, *Umbilicaria* spp.) after leaching, neutralizing with ash, or prolonged cooking. Most cannot be eaten, because they contain bitter acids and complex carbohydrates that are not easily broken down by the human digestive system. Eating raw, unprocessed lichens, even edible species, can cause stomach cramps and discomfort. (Deer, caribou, reindeer and other ungulates are much better able to digest lichens.)

Poisonous Lichens

Some lichens are quite toxic, due to the presence of usnic or vulpinic acid or other lichen substances. One example is Wolf Lichen (*Letharia vulpina*), a bright greenish yellow, finely branching fruticose type growing on trunks and dead branches of coniferous trees in dry areas of northwestern North America. In Scandinavia, it was powdered, mixed with ground glass, and sprinkled on meat, as a wolf poison. Another notably poisonous species is a ground lichen

Fig. 27. Wolf Lichen (*Letharia vulpina*), a poisonous lichen. R. & N. TURNER

(*Parmelia molliuscula*), a gray-green, irregularly dissected, flattened type described in detail by Kingsbury (1964). It grows from Nebraska to North Dakota and west to the Rocky Mountains and has been responsible for range poisoning of sheep and cattle, sometimes causing severe paralysis and death. Poisoning usually occurs in winter when other forage is scarce. Another lichen, *Bryoria tortuosa*, a look-alike relative of the hair-like Black Tree Lichen, contains much higher concentrations of poisonous vulpinic acid and is potentially toxic.

Lichens and Skin Irritation

Some lichens occasionally cause dermatitis, known as "wood-cutter's eczema." This ailment has been traced to allergic reactions of people coming in contact with lichens containing usnic acid and various other substances.

Cautionary Note on Eating Lichens

Because of the difficulty in identifying and distinguishing various lichens, their doubtful digestibility except with special preparation, and their potential toxicity, lichens should not be consumed by the uninformed. Their nutritive value is rather low, although they have been known to sustain life in emergency situations.

References for Lichens; Chapter III: Kingsbury 1964; Mitchell and Rook 1979; Richardson 1975; Smith 1921; Turner 1977.

FERNS AND FERN RELATIVES

QUICK CHECK Bracken Fern is potentially toxic and carcinogenic; Horsetails are toxic to livestock; some other ferns (Male Fern, Jimmy Fern, Sensitive Fern) are poisonous to livestock if consumed in quantity; most ferns are harmless.

Some Harmful Ferns

Several species of ferns and their relatives are known to be poisonous to humans and animals. Prominent among these are Bracken Fern and the Horsetails, both of which are treated here in detail. Others include Male Fern (*Dryopteris filix-mas*), Jimmy Fern (*Notholaena sinuata* var. *cochisensis*), and Sensitive Fern (*Onoclea sensibilis*).

The widely distributed Male Fern, used medicinally as a deworming agent, is known to contain thiaminase (see under Bracken Fern) and another potentially harmful substance called filicin. Jimmy Fern, an erect, evergreen perennial of dry, rocky hills in the south central United States and Mexico, contains an unknown toxin, apparently restricted to the one variety mentioned, that affects range stock, especially sheep.

Sensitive Fern, a perennial with green, broadly triangular vegetative fronds and deep brown spore-bearing fronds, grows in open woods and thickets throughout eastern North America. It is a suspected cause of nervous disorder in horses.

Bracken Fern, or Brake
(Hay-scented Fern Family)

Pteridium aquilinum
(Dennstaedtiaceae)

Fig. 28. Bracken Fern (*Pteridium aquilinum*), mature frond about 1 m (3 ft) high . R. & N. TURNER

QUICK CHECK Large, coarse fern of open fields and woodlands, with long-stemmed fronds of broadly triangular outline arising singly from perennial rhizomes. Poisonous to livestock; potentially carcinogenic. Eating Bracken fiddleheads not recommended.

DESCRIPTION Bracken is a common fern, usually 1–2 m (3–6 ft) high, growing from long, blackish, branching horizontal rhizomes. The young shoots (fiddleheads) are brownish and scaly, rolled tightly inwards from the tip, and bent over at top like a shepherd's crook. The frond stalks are tall, stiff, and light-brown, and the blades (leafy part) are coarse, broadly triangular, and usually three times divided, the ultimate segments oblong. The leaflet margins are inrolled, and spore-bearing structures, when present, are inconspicuous and borne on the undersides of the margins. Fronds are deciduous.

OCCURRENCE Distribution of this fern is worldwide. It grows in open fields, meadows, and woods, and is often present in large quantities. Many varieties are distinguished, including a number in North America.

TOXICITY Bracken contains a number of toxic constituents, apparently present throughout the plant. These include: a cyanide-producing glycoside (prunasin); an enzyme, thiaminase; and at least two carcinogens, quercetin and kaempferol, and an unidentified radiation-mimicking substance, apparently also mutagenic and carcinogenic.

Not all Bracken populations contain prunasin; it can be detected in the shoots by the presence of a strong, bitter "almond" odor. Thiaminase in Bracken, highest in the rhizomes and fiddleheads, can induce fatal thiamine deficiency in livestock by destroying thiamine (Vitamin B_1) reserves, and could be harmful to humans eating large quantities of the fiddleheads. The carcinogens, quercetin and kaempferol, are flavonols that have caused intestinal and bladder cancers in rats.

These toxins, especially the carcinogens, have serious implications for humans, particularly because young Bracken fiddleheads have been used as a vegetable in some parts of the world, especially Japan, and are still occasionally recommended as a "wild food" in North America. Human consumption of Bracken has been suggested as a cause of the high incidence of stomach cancer in Japan. The fiddleheads contain high concentrations of carcinogenic agents. Furthermore, humans can be exposed to the carcinogens indirectly, through drinking milk from cows grazing in Bracken-infested pastures.

TREATMENT Acute Bracken poisoning in humans is unlikely; the main danger is the long term one of possible cancer and tissue damage. Animal poisoning and its treatment is discussed in detail by Cooper and Johnson (1984). Prompt removal of the Bracken from grazing areas, or of the animals from Bracken-infested fields is the best prevention. Milking cows should not be allowed to graze Bracken at all.

NOTES The risks to humans of eating Bracken fiddleheads are potentially serious, but have not yet been fully established. However, because the plant's toxicity to livestock has been well documented, its safety for humans is at best questionable, and it should no longer be used. Native people in western North America traditionally ate roasted Bracken rhizomes, but, perhaps fortunately, these are hardly ever used at present.

Bracken fiddleheads should not be confused with the fiddleheads of Ostrich Fern (*Matteuccia struthiopteris*), which are commonly gathered from the wild in Canada and the eastern United States and are commercially harvested in the Maritimes. These are not carcinogenic, and are safe for human consumption.

Horsetails, or Scouring Rushes
(Horsetail Family)

Equisetum arvense and other
   Equisetum species
(Equisetaceae)

Fig. 29. Common Horsetail (*Equisetum arvense*).
R. & N. TURNER

QUICK CHECK Herbaceous non-flowering perennials with rush-like, jointed, green, ridged stalks (and branches if present), scratchy to the touch, usually 30–60 cm (1–2 ft) tall; leaves reduced to papery sheathing rings at the joints. Potential cause of thiamine deficiency when consumed in quantity or over prolonged periods.

DESCRIPTION There are many species of Horsetail in North America. One of the most prevalent is Common Horsetail (*Equisetum arvense*), a branching, bright-green plant, up to 60 cm (2 ft) high with stems produced annually from underground rhizomes. The stems are jointed, hollow, and rough-textured, with longitudinal ridges. The branches are slender and numerous, in whorls from the joints. The leaves are small and papery, forming a sheathing ring at each joint. A spore-bearing "cone" is produced at the end of a separate, whitish, non-branching stalk in early spring. Other species are unbranched, and some have dark green, perennial stems. In some, the spore-bearing structure grows at the tip of the green, vegetative shoot.

OCCURRENCE Common Horsetail is widespread in North America, growing in a variety of habitats, from damp, swampy areas to dry, sandy locations in woodlands, pastures, and cultivated land. Various other species occur in different regions, mostly in damp soil.

TOXICITY The main toxic compound in Horsetail is thiaminase, the same enzyme found in Bracken Fern, which destroys thiamine (Vitamin B1) in the body. The plants also contain a saponin, several flavone glycosides, and silica. Horsetail is sometimes used in herbal medicine as a diuretic and astringent, but according to Tyler (1987), its effectiveness is limited, and it should be used sparingly. The young shoots of some species are considered edible, but should not be eaten in quantity.

TREATMENT Acute poisoning in humans from Horsetail is unlikely. However, extended use of herbal preparations may result in thiamine deficiency, which could be treated by administering Vitamin B1 or yeast, or by intravenous injection of thiamine. With early diagnosis and proper nutrition, rapid recovery is possible.

NOTES The silica in Horsetails makes them scratchy to the touch, and useful as abrasives and pot scrubbers. The young shoots of Common and Giant Horsetails (*Equisetum arvense* and *E. telmateia*) were eaten as a spring-time vegetable by Native peoples of the Northwest Coast, apparently without detrimental effects, but the mature plants should not be eaten.

References for Ferns and Fern-Allies; Chapter III: Cooper and Johnson 1984; Kingsbury 1964; Hill 1986; Miller 1973; Turner 1975; Tyler 1987.

CONIFEROUS TREES

Evergreen (needled or scaled) trees, mostly cone-bearing, are known to contain a wide variety of complex chemical substances in their foliage, bark, and pitch. The "Christmas tree" aroma of Pines and Firs, and the characteristic scent of Junipers and their relatives indicates the presence of these compounds. Most of these trees would be too strong-tasting and unpalatable to eat, but many can be used safely as flavourings or to make beverage and medicinal teas, as long as these are taken in moderation and in low concentrations. Exceptions are the Yews (*Taxus* spp.), which are highly toxic and are treated in detail in Chapter IV (p. 162), and Ponderosa Pine (*Pinus ponderosa*), a tree of dry western forests, with long needles usually in clusters of three. The inner bark and seeds of this pine were eaten by Native Indians, but they knew that pregnant women should not chew on the buds or needles because it would cause a miscarriage (Turner et al. 1980). Eating the foliage of this pine is known to cause miscarriages and stillbirths in cattle and other livestock. Other pines, such as Loblolly Pine (*Pinus taeda*) of the southeastern United States, should also be regarded with caution (Kingsbury 1964).

Additionally, some trees of the cypress family (Cupressaceae)— "Cedars," Cypresses, and Junipers (including *Thuja* spp., *Cupressus*

Fig. 30. Ponderosa Pine (*Pinus ponderosa*), showing young cones. R. & N. TURNER

spp., *Chamaecyparis* spp. and *Juniperus* spp.)—have been implicated in some cases of illness and poisoning, and should be regarded with caution. Juniper "berries" have been widely used as a flavoring and herbal medicine, but are known to cause kidney irritation, uterine contractions, and possible miscarriage in pregnant women if taken in quantity (Tyler 1987).

BROAD-LEAVED TREES AND TALL SHRUBS

Buckeyes: California Buckeye, Ohio Buckeye, Sweet Buckeye, Red Buckeye
(Horse Chestnut Family)

Aesculus californica, A. glabra, A. octandra, A. pavia
(Hippocastanaceae)

Fig. 31. Buckeye (*Aesculus octandra*). MARY W. FERGUSON

QUICK CHECK Trees or shrubs with palmately compound leaves, erect flower clusters, and large, glossy brown seeds within a spiny capsule. Entire plants poisonous; children sometimes seriously poisoned from eating the seeds or drinking "tea" from the leaves. Buckeye honey also poisonous.

DESCRIPTION The Buckeyes are relatives of the Horse Chestnut, and are similar to this ornamental tree. They are deciduous shrubs or small trees, the leaves opposite and palmately compound, with 5–7 toothed leaflets. The flowers are red to yellowish, and numerous, in large, erect clusters. The fruits are 3-parted capsules with spiny, leathery, greenish to brownish husks splitting open at maturity to reveal 1–6 large, glossy brown seeds, each bearing a large, pale scar or "buckeye."

OCCURRENCE There are several species of Buckeye occurring in North America (see also Horse Chestnut, p. 156). California Buckeye grows in the dry hills and canyons of the southern Pacific Coast. Ohio Buckeye grows in rich woods from Pennsylvania to Nebraska, and south to Texas and Alabama. Sweet Buckeye is found in mountain woodlands from Pennsylvania to Iowa and southward, and Red Buckeye grows in valleys of the southeastern and south-central states. The Buckeyes are occasionally grown as garden ornamentals.

TOXICITY The main toxin in Buckeyes and Horse Chestnut is a saponin glycoside, aesculin, found in the leaves, shoots, bark, flowers, and seeds. All four species listed are toxic to humans and animals. Buckeye honey can also be poisonous.

Children are sometimes poisoned from eating the nuts or making "tea" from the leaves or twigs. Poisoning symptoms include weakness, loss of coordination, vomiting, twitching, dilated pupils, sluggishness, excitability, and paralysis.

TREATMENT Induce vomiting, or perform gastric lavage; supportive therapy for symptoms; maintain fluid and electrolyte balance.

NOTES The nuts of some Buckeye species were a traditional food of Native peoples, but had to be specially prepared to reduce the toxic effects (Appendix 2). Buckeye has been used as a fish poison.

Cherries, Wild: Choke Cherry, Black Cherry, Bitter Cherry, Pin Cherry
(Rose Family)

Prunus virginiana, P. serotina, P. emarginata, P. pensylvanica, and other *Prunus* species
(Rosaceae)

Fig. 32. Choke Cherry (*Prunus virginiana*); leaves, bark and seeds toxic. R. & N. TURNER

QUICK CHECK Trees or shrubs with simple, usually deciduous leaves, clustered white or pinkish flowers, small cherry-like fruits. Leaves, shoots, bark, and seed kernels contain cyanide-producing com-

Fig. 33. Bitter Cherry (*Prunus emarginata*);
leaves, bark and seeds toxic. R. & N. TURNER

pound; potentially fatal to children swallowing large numbers of
seeds when they eat the fruit.

DESCRIPTION There are several kinds of wild cherries in North America,
including some garden escapes. Those listed are trees or tall (or occasionally low) shrubs. The bark is rough and dark, to smooth and gray
or reddish, often peeling off in horizontal strips. The leaves,
deciduous in most species, are simple and alternate, often with
toothed edges. The flowers are white or pinkish, and 5-petalled, in
elongated or umbrella-like clusters. The cherry-like fruits are red to
blackish, with a fleshy, usually edible, layer covering a single, hard
stone.

Choke Cherry is a shrub or small tree with small flowers in elongated clusters (racemes), and red, purplish, or blackish fruit. Black
Cherry is a tree, at maturity 15 m (50 ft) or more tall and 1 m (3 ft)
across, with flowers in elongated clusters, and black fruit. Bitter
Cherry and Pin Cherry are similar, both shrubs or small to medium
trees with flowers in small, umbrella-like clusters (umbels) and small,
red fruit.

OCCURRENCE Wild Cherries are found in woods and thickets, often near
water, throughout North America except in the far north. Choke
Cherry is widespread, with varieties occurring from coast to coast in
temperate areas. Black Cherry occurs in eastern Canada and the
United States south to Florida and west to Texas. Bitter Cherry is
found along the Pacific Coast east to the Rocky Mountains, and Pin
Cherry from eastern British Columbia to the Atlantic coast, ranging
into the central and northeastern United States.

TOXICITY Although most Wild Cherries have fruit that is palatable and
pleasant to eat when ripe, several species are known to contain

dangerous levels of a cyanide-producing compound—a cyanogenic glycoside called amygdalin—in their leaves, twigs, bark, and seeds (stones). Another related compound, prunasin, has been isolated from Black Cherry (*Prunus serotina*). These compounds have been responsible for much livestock loss, as well as occasional illness and fatalities in humans. There is considerable variation in level of cyanide production among species and in different parts or growth stages. The highest concentrations of cyanide are produced in the largest, most succulent leaves on vigorous shoots. Wilting leaves usually produce larger amounts than fresh leaves.

Symptoms of cyanide poisoning include: initial rapid breathing followed by slow, difficult breathing, anxiety, excitement, confusion, dizziness, headache, vomiting (often with a pronounced bitter almond smell on the breath), initial high blood pressure and slow heart beat followed by low blood pressure and rapid heart beat, gasping, staggering, paralysis, and prostration. Convulsions, coma, and death may occur very suddenly, within a matter of minutes following very large doses.

Children are sometimes fatally poisoned by eating Wild Cherries and swallowing the cyanide-producing stones; these should always been discarded before the fruits are eaten.

TREATMENT If patient does not show any symptoms, observe closely for 2 hours. If symptoms appear, treat for cyanide poisoning (p. 12); if available use Lilly Cyanide Antidote Kit; oxygen and artificial respiration may be required, then initial inhalation of amyl nitrite to dilate blood vessels, followed by i.v. sodium nitrite, then i.v. sodium thiosulfate, to combine with hydrocyanic acid forming non-toxic thiocyanate which is readily excreted; in severe cases there may be no time to first induce vomiting or perform gastric lavage; this should be done after initial treatment; follow with supportive therapy for symptoms.

NOTES Domesticated Cherries, Plums, Apples, Pears, Peaches, Apricots, and Almonds contain the same cyanide-producing compound in their seeds, leaves, and bark as the Wild Cherries. Other wild plants in the rose family known to be cyanide-producing and hence toxic are Mountain Mahogany (*Cercocarpus* spp.) and American Cherry Laurel (*Prunus caroliniana*). Wild Plums are also toxic, but less likely to cause poisoning in humans because their stones are larger and are seldom swallowed. Serviceberry, or Saskatoon and its relatives (*Amelanchier* spp.) have edible fruit, but their leaves and twigs may produce toxic amounts of cyanide.

Coyotillo and Its Relatives
(Buckthorn Family)

Karwinskia humboldtiana and
related species
(Rhamnaceae)

QUICK CHECK Simple-leaved shrub or small tree of dry hills, canyons and
valleys of the far Southwest; brownish black "berries" about ½ in.
across. Eating "berries" causes delayed paralysis in humans and
animals; foliage also poisonous.

DESCRIPTION This is a shrub or small tree 1–6 m (3–20 ft) tall. The leaves,
mostly opposite, are stalked and elliptical to oval, with smooth or
wavy, slightly recurving margins and distinct straight veins. The
small, greenish flowers are borne in clusters at the leaf axils. The fruit
is an ovoid, berry-like drupe (fleshy outer covering, with single
stone) about 1.2 cm (0.5 in.) across.

OCCURRENCE Dry, gravelly hills, canyons, and valleys in southwestern
Texas, New Mexico, Arizona, southern California, and Mexico.

TOXICITY The fleshy fruits have long been known to be poisonous to
humans and livestock, including goats, sheep, cattle, hogs, and fowl.
The foliage is also toxic and may be fatal to browsing livestock. Eating
the fruits can cause severe paralysis, which may be delayed for a few
days to several weeks without any visible signs of poisoning in the lag
period. Then weakness, lack of coordination, and finally complete
prostration may ensue. Recovery may take many weeks. The toxin is
an incompletely identified quinone.

TREATMENT Induce vomiting or perform gastric lavage if fruits have been
eaten; general supportive therapy for symptoms with delayed
paralysis.

NOTES Children should be kept away from the "berries." The delayed
action of the paralysis caused by eating the fruit can make detection of
the poisoning source very difficult. The seeds are the most toxic part
of the plant.

**Elderberries: Red-berried,
Blue-berried, and Common
Elders**
(Honeysuckle Family)

*Sambucus racemosa, S. cerulea,
   S. canadensis,* and related
   Sambucus species
(Caprifoliaceae)

Fig. 34. Red-berried Elder (*Sambucus racemosa*);
leaves and bark toxic; raw fruits may cause nausea.
R. & N. TURNER

QUICK CHECK Tall deciduous bushes, with opposite, pinnately compound leaves, and small red, blueish or purple-black berries in flat-
topped or pyramidal clusters. Leaves, stems, bark, and roots strongly
purgative and cyanide-producing; uncooked berries may cause
nausea in humans.

DESCRIPTION These are coarse bushes, up to 4 m (12 ft) high, with stout,
pithy, grayish-barked stems. The leaves are opposite and pinnately
compound, with lance-shaped, toothed leaflets. The small, numerous
flowers are whitish or creamy, in dense flat-topped or pyramidal
clusters. The berries are small, but produced in large, conspicuous
masses. Depending on the species, the fruits range in color from
bright-red to grayish blue to purple-black. Young growths may be
confused with Water Hemlock (*Cicuta* spp.), but may be distinguished by their opposite leaves and absence of tuberous roots.

OCCURRENCE Moist, rich soils in woodlands throughout much of North
America. Red-berried Elder (*S. racemosa*) occurs from Alaska south to
California and New Mexico, and east to the Maritimes in Canada.
Blue-berried Elder (*S. cerulea*) is found from British Columbia west to
Montana and south to New Mexico. Common Elder (*S. canadensis*)
grows in eastern North America from Manitoba to the Maritimes and
south to Georgia and Louisiana. A fourth species, Gulf Elder (*S.
simpsonii*), occurs from Florida to Louisiana.

TOXICITY The roots, stems, bark, and leaves, and to a much lesser extent,
the flowers and unripe fruits, contain a poisonous alkaloid and
cyanogenic glycoside causing nausea, vomiting and diarrhea. People
are sometimes poisoned by drinking medicinal "tea" made from

Elder leaves or branches. Children have been poisoned by using hollowed elder stems for popguns and peashooters.

TREATMENT If symptoms are evident, treat for cyanide poisoning (see p. 12; also under Cherries, Wild in this section—p. 156). Induce vomiting, or perform gastric lavage after cyanide treatment has been carried out.

NOTES Elderberry stems and roots are used medicinally as an emetic and purgative by a number of North American Indian groups.

Elder flowers and fruits are edible, and the latter are commonly used to make wine and jelly. However, there have been reports of the raw fruits of Red-berried Elder causing nausea (see Appendix 2).

Kentucky Coffee Tree
(Bean, or Legume Family)

Gymnocladus dioica
(Fabaceae, or Leguminosae)

Fig. 35. Kentucky Coffee Tree (*Gymnocladus dioica*)
R. & N. TURNER

QUICK CHECK Large, much-branched tree of the eastern United States, with large, divided leaves and large, brown, hard "bean" pods. Seeds poisonous to humans but seldom fatal; foliage and shoots highly toxic.

DESCRIPTION A large, rough-barked forest tree up to 30 m (80 ft) or more tall, with short trunk and many major branches. The leaves are alternate, twice pinnately divided, and up to 1 m (3 ft) long, with oval-shaped, smooth-margined leaflets. The flowers, produced in May, are whitish, in elongated terminal clusters. The fruit is a hard, flat, reddish brown "bean" pod up to 15 cm (6 in.) long and about 2.5 cm (1 in.) broad, containing 4–7 hard, flat seeds with sticky pulp between them. The fresh pods exude yellow resin when broken.

OCCURRENCE Rich, moist woods from southern Ontario to Alabama and Oklahoma; occasionally planted elsewhere as a lawn or street tree.

TOXICITY The main toxic substance in this tree is reported to be a quinolizidine alkaloid, cytisine. Humans may be poisoned by eating the seeds or pulp between the seeds, although the roasted seeds were

Fig. 36. Kentucky Coffee Tree, pods with seeds.
WALTER H. LEWIS

used by early settlers as a coffee substitute. Livestock have been fatally poisoned from eating the foliage or young shoots. Symptoms of poisoning include intense irritation of the digestive tract, diarrhea, vomiting, congestion of mucous membranes, irregular pulse, and coma. Death may occur within a day after appearance of symptoms.

TREATMENT Induce vomiting, or perform gastric lavage; activated charcoal, and supportive measures for symptoms, including artificial respiration and oxygen.

NOTES The fruits of this tree may be confused with those of the more common Honey Locust (*Gleditsia triacanthos*), whose seed pulp is sweet and edible. The pod of this latter tree is much longer (20–30 cm; 8–15 in.), often twisted or curved, with thinner walls. The Honey Locust is also usually thorny. [Note: Do not confuse *Gleditsia* with Black Locust, or False Acacia (*Robinia pseudo-acacia*) (p. 148), another poisonous-fruited tree of the bean or legume family, which is also sometimes called "Honey Locust."]

Keep children away from the pods of Kentucky Coffee Tree. Livestock should be kept away from the felled trees, cut branches, and sprouts from cut stumps.

Laurels: Dwarf Laurel, Mountain Laurel, and Swamp Laurel
(Heather Family)

Kalmia angustifolia, K. latifolia,
   and *K. polifolia*
(Ericaceae)

Fig. 37. Dwarf Laurel (*Kalmia angustifolia*).
MARY W. FERGUSON

QUICK CHECK Low shrubs to small trees of rocky or sandy acidic soils or peat bogs with dark green, evergreen leaves and whitish to pink or

Fig. 38. Mountain Laurel (*Kalmia latifolia*).
MARY W. FERGUSON

Fig. 39. Swamp Laurel (*Kalmia polifolia*).
MARY W. FERGUSON

deep-red, clustered, bell-shaped to urn-shaped flowers. Leaves and flowers highly toxic, sometimes fatal, to humans and browsing animals. Flower nectar and honey also toxic.

DESCRIPTION The Laurels all have smooth, evergreen leaves and showy, clustered, 5-pointed, ornate flowers with the petals fused together, and dry, capsulate, many-seeded fruits.

Dwarf Laurel, or Lambkill, is an open, woody shrub, 0.3–1.3 m (1–4 ft) tall. The leaves are opposite, or whorled in threes, and elongated, 2.5–6.5 cm (1–2.5 in.) long. The branches are upright and the flowers rose or crimson, widely bell-shaped, and produced in lateral clusters.

Mountain Laurel is a shrub or small tree 2–10 m (6–30 ft) high, with alternate leaves up to 10 cm (4 in.) long. The flowers are white to rose with purple markings, and borne in terminal clusters.

Swamp Laurel, or Bog Laurel, is a small shrub usually 30–60 cm (1–2 ft) high. The leaves are opposite and shiny, whitish beneath, and usually less than 2.5 cm (1 in.) long. The flowers, rose to purple, are similar to those of Dwarf Laurel but smaller, and in terminal clusters. Only a small-leaved variety, *Kalmia polifolia* var. *microphylla*, has been reported to be toxic.

OCCURRENCE Dwarf Laurel is found in old pastures and meadows throughout northeastern North America. Mountain Laurel grows in moist woods and clearings and along streams from eastern Canada southward in the Appalachian Mountains and piedmont, and infrequently in the eastern coastal plain of the United States. Swamp Laurel occurs in peat bogs and wet meadows throughout northern North America; var. *microphylla* occurs in the Rocky Mountains from northern California to Alaska.

TOXICITY Mountain Laurel has caused human deaths. All three species have produced livestock deaths, especially of sheep and cattle. Poisoning is due to a resin, andromedotoxin, and to a glucoside of hydroquinone, arbutin. There is an initial burning of the lips, mouth, and throat with ingestion of the plants, followed up to 6 hours later by salivation, nausea, severe vomiting accompanied by abdominal pain and watering of the mouth, eyes, and nose, loss of appetite, repeated swallowing, headache, low blood pressure, and drowsiness, with to convulsions, weakness, difficult breathing, and progressive paralysis of the limbs, followed by coma and death in the most severe cases.

TREATMENT Induce vomiting, or perform gastric lavage; follow with activated charcoal and saline cathartic; maintain fluid and electrolyte balance; monitor heart, blood pressure, and breathing. Atropine may be used to stimulate the heart, dopamine to treat low blood pressure; oxygen and assisted breathing as required; convulsions or excitation may be controlled with i.v. diazepam.

NOTES Children have been poisoned by sucking the flowers of Mountain Laurel and making "tea" from the leaves. Even honey from the flowers is poisonous (see Appendix 5). Laurels are usually avoided by livestock, but poisoning can occur especially in late spring and fall, when no other green foliage is available. At the Washington Zoo, a valuable monkey and angora goats were poisoned at different times by being fed Laurel flowers and leaves by visitors.

Several other plants in the heather family contain the same poisonous compounds as the Laurels. These include: Rhododendrons, Pieris, False Azalea and Labrador-tea. Laurels are sometimes grown as garden ornamentals.

Mescal Bean, or Frijolito
(Bean, or Legume Family)

Sophora secundiflora
(Fabaceae, or Leguminosae)

Fig. 40. Mescal Bean (*Sophora secundiflora*), showing bright red seeds. R. & N. TURNER

QUICK CHECK Evergreen shrub of dry areas of southwestern Texas; pinnately compound, often hairy leaves; purplish, pea-like, clustered flowers; large, jointed, cylindrical fruiting pods. Seeds hallucinogenic and highly toxic; potentially fatal in small quantities to children; entire plant toxic to livestock but rarely fatal.

DESCRIPTION This evergreen shrub or small tree, to 10 m (35 ft) tall, has leaves that are stalked and alternate, 10–15 cm (4–6 in.) long, and pinnately compound with 7–13 leathery, smooth-edged, oblong leaflets. The leaves may be covered with dense, whitish hairs. The pea-like flowers are violet-blue and very fragrant, produced in large, one-sided, terminal clusters. The fruits are woody, jointed, cylindrical pods each 2.5–18 cm (1–7 in.) long, containing 1–8 bright red, hard seeds which average about 13 mm (0.5 in.) long.

OCCURRENCE Mescal Bean grows naturally on dry limestone hills and canyons of southwestern Texas and New Mexico into Mexico. Widely grown in the southern United States as an ornamental.

TOXICITY This plant, like other species of the genus *Sophora*, contains poisonous quinolizidine alkaloids including cytisine, which is also present in Laburnum. The seeds are most commonly implicated in human poisoning. Symptoms include nausea, vomiting, diarrhea, excitement, delirium, convulsions, coma, and occasionally death through respiratory failure. One seed, thoroughly chewed, is said to be enough to kill a child.

TREATMENT Induce vomiting, or perform gastric lavage; treat for symptoms; general supportive measures. (see also under Laburnum—p. 158).

NOTES Mescal Bean is known to be hallucinogenic and has long been used medicinally and ritually by Southwest Indians.

Oaks
(Beech Family)

Quercus spp.
(Fagaceae)

Fig. 41. Garry Oak (*Quercus garryana*) leaves and acorns. R. & N. TURNER

QUICK CHECK Usually deciduous (some evergreen) trees or shrubs with alternate, simple, often lobed or toothed leaves, and capped fruits (acorns). Young shoots, foliage, and some acorns poisonous. Acorns of some species eaten by humans, but only after special treatment to remove toxins.

DESCRIPTION Oaks are very common and probably recognized by most people. They are a large, variable group with some 60 native tree-size species in North America. There are also many hybrids and varieties, shrubby species, and introduced ornamentals. Oak leaves are alternate, simple, pinnately veined, and variable in size and shape. The flowers are greenish or yellowish, solitary or clustered, often appearing before the leaves. The fruits are spherical or elongated acorns partially enclosed by a scaly cup. Native Oaks can be subdivided into three main groups: the White Oaks, with deciduous leaves that are usually deeply lobed with rounded edges; the Red Oaks, with deciduous, usually sharply lobed, often bristle-tipped leaves; and the Live Oaks, with leaves that are thick, leathery, often spiny-margined, and remain green through the winter.

OCCURRENCE Open, dry to moist deciduous woods throughout North America.

TOXICITY The acorns, leaves, and young shoots of many different Oak species are toxic, reportedly due to the presence of tannins of the gallotannin class. There are, fortunately, no reported human fatalities from Oaks. Acorns, especially of the White Oak group, were an important traditional food for Native peoples of North America. However, the kernels had to be carefully processed to remove the astringent, toxic principles. Since taste is an important factor in the edibility of acorns, poisoning was probably avoided by the unpalatability of those having a high tannin content. Nevertheless, the authors are aware of several people who consumed raw, untreated acorns on an experimental basis, resulting in painful and lingering irritation of the mouth and throat. There is also a danger of young children being poisoned from acorns. In severe cases, the tannins can cause inflammation, irritation, and hemorrhaging of the intestinal walls and degeneration of the liver and kidneys.

TREATMENT Induce vomiting and use demulcents to coat the digestive tract; treat for symptoms.

NOTES There are innumerable records in North America and Europe of livestock loss from Oak shoots, leaves, and acorns. Cattle are particularly susceptible to Oak poisoning, and sheep and goats are also affected. Poisoning usually occurs in spring when new foliage is eaten, in the fall of heavy acorn crop years, and during drought years when other forage is scarce.

Poison Sumac, Swamp Sumac, or Poison Elder
(Cashew Family)

Toxicodendron vernix (syn. *Rhus vernix*)
(Anacardiaceae)

Fig. 42. Poison Sumac (*Toxicodendron vernix*) (right) and Poison Ivy (*T. radicans*) (left). J. C. RAULSTON

QUICK CHECK Low shrub to small tree of bogs and swamps, with deciduous, alternate, pinnately compound leaves. Entire plant causes severe allergenic skin irritation.

DESCRIPTION A low, much-branched shrub, or, in crowded stands, a single-trunked tree up to 8 m (25 ft) tall. The bark is smooth and grayish. The leaves are deciduous, alternate and pinnately divided into 7–13 leaflets, each short-stalked, pointed, and smooth-edged. The leaf stalks are reddish. The flowers are tiny, greenish yellow, and inconspicuous; the fruits are small, dry, whitish to creamy yellow berry-like drupes, in hanging clusters that may persist over the winter.

OCCURRENCE Swamps, wet woods and around boggy ponds from southern Ontario and Quebec south to eastern Texas and Florida; predominantly east of the Mississippi River.

TOXICITY The sap of the roots, stems, leaves, pollen, flowers and fruits contains a potent skin irritant, released by contact with the plant (especially with bruised portions), and affecting many people who have an allergic sensitivity to it. The toxic components have not been fully identified, but are similar to those of Poison Ivy. Danger of poisoning is greatest in spring and summer when the sap is produced abundantly. Symptoms are the same as for Poison Ivy (see p. 87).

TREATMENT See under Poison Ivy (p. 87).

NOTES Other, non-irritant shrubby Sumacs (*Rhus* spp.) can be distinguished from Poison Sumac by having toothed leaflets, and flowers and fruits in dense, terminal, erect clusters. They also grow in drier habitats. Elderberries (*Sambucus* spp.) and Ashes (*Fraxinus* spp.) may also be confused with Poison Sumac due to their pinnately compound leaves, but their leaves are opposite, not alternate.

Another irritant plant of the same family, Poisonwood (*Metopium toxiferum*), is a tall shrub or small tree found in Monroe and Dade counties in Florida. It has compound, 5–7 parted leaves and fleshy fruits about 2.5 cm (1 in.) long. It is presumed to contain the same poisonous compounds as Poison Sumac and Poison Ivy.

MEDIUM-TO-LOW SHRUBS: EVERGREEN AND DECIDUOUS

Broom, or Scotch Broom
(Bean, or Legume Family)

Cytisus scoparius (syn.
  Sarothamnus scoparius)
(Fabaceae, or Leguminosae)

Fig. 43. Broom (*Cytisus scoparius*). R. & N. TURNER

QUICK CHECK Upright, bushy shrub of open areas; twigs green and angled; flowers pea-like, bright yellow; fruits flat pea-like pods. Potentially poisonous to humans and livestock, but no fatalities reported.

DESCRIPTION Broom is a stiffly upright, much branching shrub up to 2 m (6 ft) or more high, with angled, evergreen twigs. The leaves are small and deciduous, usually 3-parted. The flowers are numerous, bright yellow and pea-like (sometimes whitish or marked with dark red); often more-or-less covering the bushes. The fruit is a flat pod, green with whitish, silky hairs along the margins when young, turning blackish at maturity and snapping open explosively, the two halves twisting backwards to release the small, brownish seeds.

OCCURRENCE This weedy shrub was introduced from Europe and is now widely naturalized in North America. It is common on dry hillsides, roadsides, wastelands, and shorelines on both the Atlantic and Pacific coasts.

TOXICITY Broom contains several toxic quinolizidine alkaloids, including sparteine, isosparteine, cytisine, and lupinidine. These compounds can depress the heart and nervous system, sometimes paralyzing the

Fig. 44. Broom, unripe seed pods.
R. & N. TURNER

motor nerve endings. Fortunately, the alkaloids are present in relatively small amounts, and only a few cases of livestock poisoning from Broom have been reported. Still, it should be regarded as potentially dangerous to humans as well as animals. Symptoms would be expected to be similar to those of Lupine poisoning: nausea, vomiting, dizziness, headache, and abdominal pain.

TREATMENT In case of suspected Broom poisoning, treat for symptoms and apply general supportive measures (see also under Laburnum— p. 158).

NOTES Broom seeds have been used as a coffee substitute, and Broom flowers have been pickled, and used to make wine. However, extreme caution is advised. There is a potential for children to be poisoned from eating the pods and seeds, which resemble small peas.

Poison Ivy
(Cashew Family)

Toxicodendron radicans (syn. *Rhus radicans;* includes *T. rydbergii,* sometimes treated as a separate species)
(Anacardiaceae)

Fig. 45. Poison Ivy (*Toxicodendron radicans*), in flower.
MARY W. FERGUSON

QUICK CHECK Low shrub (usual western form) or woody vine (usual eastern form) with 3-parted, pointed leaves, inconspicuous flowers, and whitish, clustered berrylike fruits. Entire plant is allergenic, causing severe and continuing skin irritation on contact for most people.

DESCRIPTION This woody plant is quite variable. It is typically a trailing or climbing vine in eastern North America, and a low, bushy shrub in the West. The leaves are alternate, 3-parted, often drooping, with shiny, oval, pointed leaflets having smooth, toothed, or lobed edges. The flowers are small and inconspicuous, in hanging clusters, and the fruit is a smooth, whitish, berry-like drupe. The seeds are white and longitudinally grooved.

OCCURRENCE Widespread in North America, with several varieties or forms; found in sandy, gravelly, or loamy soil in disturbed places, on flood plains, along lakeshores and streambanks, and in woods throughout southern Canada and the United States, except on the Pacific Coast.

TOXICITY Poison Ivy is the best known and most widespread of a group of related species including Poison Oak and Poison Sumac, all con-

Fig. 46. Poison Ivy, fruiting. R. & N. TURNER

taining the same highly irritating, widely allergenic oleoresin, urushiol, which is composed of a mixture of compounds, the worst being a phenolic resin, 3-n-penta-decyl-catechol. All parts of the plant—roots, stems, leaves, flowers, and fruits—are potentially irritating, and even the pollen, smoke from burning the plant, or clothing and tools coming in contact with the plant, can produce symptoms.

Symptoms of Poison Ivy allergy include itching, burning, and redness of the skin. Small blisters may appear after a few hours or as long as five days after contact. Severe dermatitis, with large blisters, swelling, headache, and fever, may occur in some individuals, or after prolonged contact, and may require hospitalization. Fluid in blisters contains traces of the irritant and if the blisters are broken, they can further irritate the surrounding skin or become infected.

TREATMENT Remove contaminated clothing. Immediately wash all affected areas thoroughly with strong soap and water, then apply rubbing alcohol. Do not break blisters. Apply first aid cream with antihistamine to ease the itching. Keep contaminated clothing separate from other laundry and wash thoroughly several times. In severe cases, see physician.

NOTES Not everyone is equally sensitive to Poison Ivy and its relatives, but most people are potentially vulnerable to them sooner or later. Sensitivity is usually acquired in childhood or early adult life, and tends to decline in later life.

Poison Oak and Western Poison Oak
(Cashew Family)

Toxicodendron quercifolium (syn. *Rhus quercifolia*) and *T. diversilobum* (syn. *R. diversilobum*) (Anacardiaceae)

Fig. 47. Western Poison Oak (*Toxicodendron diversilobum*). ROBERT A. ROSS

QUICK CHECK Woody shrubs, or climbing vines (Western Poison Oak only), with 3-parted, oak-like leaves; flowers and fruits in hanging clusters; fruits yellowish or whitish and berry-like. Entire plants cause severe allergenic skin irritation on contact.

DESCRIPTION Poison Oak (*T. quercifolium*) is a woody shrub, never vine-like, whereas Western Poison Oak (*T. diversilobum*) is sometimes shrubby, but is commonly a vine up to 15 m (50 ft) tall. The leaves are alternate and 3-parted (sometimes 5-parted), the leaflets deeply toothed or lobed and resembling oak leaves. The small, greenish flowers are borne in loose, hanging clusters. The fruits are small, berry-like drupes, each containing a single seed. Western Poison Oak berries are whitish, whereas those of *T. quercifolium* are yellowish and hairy. Western Poison Oak most closely resembles Poison Ivy, and occasionally hybridizes with it.

OCCURRENCE Poison Oak (*T. quercifolium*) grows in sandy soil, of the dry barrens and oak-pine and pine woods of the Atlantic and Gulf coastal plains from southern New Jersey to northern Florida and west to eastern Texas and Kansas. Western Poison Oak (*T. diversilobum*) is mainly confined to thickets and wooded slopes of the Pacific coastal region from Puget Sound to California and Mexico, west of the Cascades and Sierra Nevada ranges, and inland along the Columbia River.

TOXICITY The sap of all parts of these species contain the same irritating oleoresin (urushiol) and its derivatives found in Poison Ivy (*T. radicans*) and Poison Sumac (*T. vernix*). Most people are more or less allergic to this substance, and on contact break out in a burning, itching rash. For details of toxicity and treatment, see under Poison Ivy (p. 87).

TREATMENT See Poison Ivy (p. 87)

NOTES According to Hardin and Arena (1974), every year nearly two million people in the United States experience irritating or painful effects from direct or indirect contact with Poison Oak, Poison Ivy, or Poison Sumac. It is estimated that seven out of every ten people are immediately allergic to these plants, and almost everyone is after prolonged or continuous exposure.

Fig. 48. Waxberry (*Symphoricarpos albus*).
R. & N. TURNER

Waxberry, or Snowberry
(Honeysuckle Family)

Symphoricarpos albus (syn.
  S. rivularis)
(Caprifoliaceae)

QUICK CHECK Low to medium deciduous shrub with rounded, opposite leaves, and white, clustered berries. Berries reported to be poisonous, especially for children.

DESCRIPTION Waxberry is an erect, much-branching, deciduous shrub up to 1 m (3 ft) or more high. The leaves are dull green, 2–5 cm (1–2 in.) long, oval-shaped, and sometimes deeply lobed. The small, pinkish white, clustered flowers are bell-shaped and hairy inside. The berries are waxy white, pea-sized, spherical, and tightly clustered, often remaining on the bushes over winter.

OCCURRENCE Dry, open woods throughout most of temperate North America from Alaska and the Northwest Territories south to California and Colorado, and east to Virginia. Sometimes grown as an ornamental in North America and Europe.

TOXICITY The berries have a long-standing reputation for being poisonous, especially for children. A reputedly toxic substance, viburnin, is found in the plant. It contains several alkaloids, including chelidonine (a narcotic characteristic of the poppy family), as well as saponins, tannins, terpenes, triglycerides, and coumarins. The toxicity of most of these compounds has not been determined.

Cases of Waxberry poisoning have been documented in children in the United States, Britain, and Poland. In Britain, a child eating only three berries experienced vomiting, slight dizziness and mild sedation. Gastrointestinal irritation, blood-stained urine, delirium, and a semi-comatose state are other reported symptoms. On Vancouver Island an elderly Saanich Indian man told one of the authors (NT) that his younger sister had died at the age of six (about 75 years ago) from eating the berries. Most Native people regard the berries as inedible and fatally poisonous if many are eaten.

TREATMENT Induce vomiting, or perform gastric lavage; treat for symptoms; general supportive measures.

NOTES Several other species of Waxberry occur in different parts of North America. All should be regarded as potentially toxic. Other members of the honeysuckle family, including the Honeysuckles (*Lonicera* spp.) and Highbush Cranberries (*Viburnum* spp.), are popularly considered to be poisonous, but no definite reports of their causing harm are to be found. Fruits of several *Viburnum* species are frequently eaten (see also under Elderberries, p. 78).

Among the Lillooet Indians of British Columbia, eating one or two Waxberries is said to be an antidote to discomfort caused by eating too much fatty food. The berries are known in several Indian languages as "ghost-berries" or "corpse-berries."

WILD VINES

NOTE: Poison Ivy (*Toxicodendron radicans*) and Western Poison Oak (*Toxicodendron diversilobum*), both of which may grow as climbing vines, are treated in the MEDIUM-TO-LOW SHRUBS section.

Manroot, Bigroot, or Wild Cucumber
(Cucumber Family)

Marah oreganus (syn.
   Echinocystis oregana)
(Cucurbitaceae)

Fig. 49. Manroot (*Marah oreganus*). R. & N. TURNER

QUICK CHECK Trailing or climbing vine with large, woody root, palmately lobed leaves, and oval, usually spiny, green fruits with large, rounded seeds. Seeds, fruits, and entire plant considered toxic, potentially fatal.

DESCRIPTION This cucumber-like plant is a herbaceous perennial vine with branching tendrils, growing from a much enlarged, woody root. The leaves are long-stalked and rather irregularly palmately lobed.

Fig. 50. Wild Balsam-Apple (*Momordica charantia*).
MARY W. FERGUSON

Fig. 51. Wild Balsam-Apple, ripe fruit. WALTER H. LEWIS

The small, greenish white flowers, in elongated clusters, bloom from April to June. The fruits are elliptical to oval, usually 3–8 cm (1–3 in.) long, green, and usually prickly, containing 2–8 large, rounded, brown seeds.

OCCURRENCE Bottomlands, fields, thickets, open hillsides, coastal bluffs, and roadsides from southern British Columbia to northern California, mostly west of the Cascade Mountains.

TOXICITY Until recently, it was not realized that this plant is fatally toxic under some circumstances. In 1986 an Oregon man went into shock, lost consciousness, and died 24 hours after consuming a homemade potion of the seeds. Within two hours of drinking at least one cup of the tea, he complained of chest pains and tightness of the chest, followed by loss of muscle control and shortness of breath. Restlessness, low blood pressure, and internal bleeding due to loss of the blood clotting function followed. Death was attributed to heart failure and internal bleeding. It is unclear why he took the drink, but possibly it was to obtain some hallucinogenic effect from the seeds. The cucumber-like fruits themselves are very bitter, but because they resemble cucumbers and children sometimes play with them, their potential toxicity is notable. The poisonous principles may be cucurbitacins, which are compounds produced by saponic glycosides known to occur widely in the cucumber family. Research on the pharmacological effects of this plant is continuing at Oregon State University (Dr. George Constantine, personal communication 1987).

TREATMENT Induce vomiting, or perform gastric lavage; treat for symptoms.

NOTES Northwestern Indians used Manroot medicinally. The seeds were eaten for kidney trouble, a decoction of the plant was drunk for venereal diseases, and the crushed roots were used as a poultice for sores of horses. The root, which is very bitter, was crushed and placed in streams to stupify fish. It is apparently rich in tannins; the Mexicans reportedly used it for tanning.

Another plant of the cucumber family reported to be poisonous is Wild Balsam-Apple (*Momordica charantia*), a small, yellow-flowered, cucumber-like vine of sandy soils and waste grounds in the coastal plain from Florida to Texas, and often cultivated in the southern Midwest. The foliage and the outer coat and seeds of the gourd-like fruits are cathartic, producing diarrhea and vomiting if eaten. The plant contains alkaloids, a resin, and a saponic glycoside yielding elaterin (a cucurbitacin). Recovery from poisoning can be lengthy.

Moonseed
(Moonseed Family)

Menispermum canadense
(Menispermaceae)

Fig. 52. Moonseed (*Menispermum canadense*).
WALTER H. LEWIS

QUICK CHECK Woody vine with grape-like leaves and fruit; fruits with single, crescent-shaped seed. Fruits highly toxic, sometimes fatal.

DESCRIPTION Moonseed is a woody, perennial, twining vine resembling grape. The leaves are alternate and palmately lobed, with 3–5 shallow smooth-edged lobes (rather than the prominent saw-toothed lobes of grape). The greenish white flowers are borne in small clusters and bloom from June through July. The fruits are globular and purplish black, in grape-like bunches, but each contains a single, large, grooved, crescent-shaped seed, rather than several to many teardrop-shaped seeds as in grapes.

OCCURRENCE This vine is found growing on other vegetation in woods, thickets and fencerows of eastern North America, from Canada south to Georgia and Oklahoma.

TOXICITY Moonseed contains isoquinoline alkaloids, including dauricine, a compound with curare-like action. The fruits, which resemble small purple grapes, are the main cause of poisoning, and when eaten in quantity can be fatal. There are reports of them causing loss of life in children mistaking them for grapes.

TREATMENT Induce vomiting, or perform gastric lavage; follow with activated charcoal and a saline cathartic; general treatment for symptoms.

NOTES Birds eat Moonseed fruits readily, apparently without harm. Contrary to popular belief, many berries and plants poisonous to humans are eaten by birds and animals. Observing wildlife eating plants is no guarantee that they can be safely eaten by people.

Nightshade, Climbing, or European Bittersweet
(Nightshade Family)

Solanum dulcamara
(Solanaceae)

Fig. 53. Climbing Nightshade (*Solanum dulcamara*).
R. & N. TURNER

QUICK CHECK Slender vine with simple leaves often basally lobed, deep purple potato-like flowers, and bright red clustered berries. Entire plant, especially unripe fruit, toxic and potentially fatal.

DESCRIPTION Slender, woody-based perennial vine up to 2 m (6 ft) or more high. The leaves are dark green, pointed, and oval to heart-shaped, often with two opposite lobes at the base. The clustered, deep purple or blueish flowers have recurving petals and bright yellow exerted stamens. They bloom throughout the summer. The berries are oval, hard and green when unripe, maturing to soft, translucent, bright red, and shiny.

OCCURRENCE This vine, introduced from Europe, is commonly found trailing over the ground or twining on other plants in damp woods, thickets, waste places, and occasionally gardens, throughout North America.

TOXICITY The entire plant contains solanine, a toxic alkaloidal glycoside also present in green potatoes, and in other members of the genus *Solanum*. Climbing Nightshade also contains a glycoside, dulcamarine, similar in its structure and effects to atropine (see Belladonna, *Atropa belladonna*—p. 183). Climbing Nightshade berries have been implicated in the fatal poisoning of children and the plants have been responsible for livestock loss in both Europe and North America. The ripe fruits are less toxic than the leaves and unripe

Fig. 54. Climbing Nightshade
fruit. R. & N. TURNER

berries, but even fully ripe berries should be considered poisonous. The actual quantity of toxins varies with light, soil, climate, and growth stage. Symptoms of poisoning include varying degrees of abdominal pain, headache, flushing and irritation of the skin and mucous membranes, and tiredness. In severe cases, vomiting, thirst, difficult breathing, restlessness, subnormal temperature, paralysis, dilated pupils, diarrhea, blood in urine, shock, extreme weakness, loss of sensation, and occasionally death, may occur.

TREATMENT Induce vomiting, or perform gastric lavage; follow with activated charcoal and a saline cathartic; treat for symptoms as they appear; maintain fluid and electrolyte balance; support respiration; paraldehyde (2–10 cc) IM is suggested.

NOTES The highest concentrations of toxic solanine are reported in the unripe fruit of *Solanum* species. In some strains of various species (see Nightshade, Black, *Solanum nigrum* and *S. americanum*—p. 125), the fully ripe berries are actually edible, but unless edibility is absolutely certain, leave the berries of all species alone. Recently, leaves of Climbing Nightshade were found by Adam Szczawinski in a restaurant salad. Since it is a common weed, it was probably gathered by mistake together with the spinach which made up most of the salad.

Peas, Wild, or Vetchlings
(Bean, or Legume Family)

Lathyrus spp.
(Fabaceae, or Leguminosae)

Fig. 55. Wild Sweet Pea (*Lathyrus latifolius*).
R. & N. TURNER

QUICK CHECK Herbaceous pea-like vines with white, yellowish, or purplish pea-like flowers and long, flat pea-like pods. Foliage and seeds
toxic and potentially fatal when consumed in quantity over an
extended period; small quantities not harmful.

DESCRIPTION There are many species of the genus *Lathyrus* to be found in
North America, some introduced from Europe, some native. All are
climbing vines with herbaceous, winged or angular stems. The leaves
are alternate and compound, divided into 2 to many elongated or
rounded leaflets and terminating in slender tendrils. The flowers are
pea-like, whitish or yellowish to shades of pink or purple, and often
quite showy. The fruits are elongated, flattened, pea-like pods, usually
containing several small pea-like seeds.

OCCURRENCE The wild peas are widespread, occurring from seaside to
mountains in a variety of habitats, from marshes and open woods to
fields and wasteland, throughout North America.

TOXICITY The foliage and especially the pea-like seeds of many species in
this genus, when consumed in quantity, can be poisonous to humans
and livestock, due to the presence of toxic amino acids. Small or even
moderate amounts have been eaten without apparent harm. Long-
term or intensive use of the seeds as food can result in a potentially
fatal disease called "lathyrism" (for further details, see under Sweet
Pea in Chapter IV (p. 217).

TREATMENT Unless large quantities are ingested, treatment is unneces-
sary. If poisoning or "lathyrism" is suspected, change diet and give
supportive therapy for symptoms.

NOTES Cooking the seeds does not eliminate their long-term toxicity for
humans. However, toxicity is caused only when large quantities are

eaten, not when relatively small amounts are consumed as part of a mixed diet. Wild *Lathyrus* species known to have seeds potentially poisonous to humans include Caley Pea (*L. hirsutus*), Wild Pea (*L. incanus*), Singletary Pea (*L. pusillus*), Chick Pea (*L. sativus*), Everlasting Pea (*L. sylvestris*), and Wild Sweet Pea (*L. latifolius*). The garden Sweet Pea (*L. odoratus*) may also cause poisoning. All species of *Lathyrus* should be regarded with caution.

Common Vetch
(Bean, or Legume Family)

Vicia sativa
(Fabaceae, or Leguminosae)

Fig. 56. Common Vetch (*Vicia sativa*). R. & N. TURNER

QUICK CHECK Small, herbaceous vine with pinnately compound leaves, purple pea-like flowers, and narrow, pointed pods. Foliage and seeds toxic and potentially fatal if large quantities are consumed.

DESCRIPTION Herbaceous vine usually under 1 m (3 ft) tall, climbing on other plants, especially grasses, by means of slender tendrils. The leaves are small and pinnately divided, with paired, oval leaflets about 1–2 cm (0.5 in.) long, and pointed stipules at the leaf bases. The small, pea-like flowers, pale pink to deep reddish purple, are borne singly or in pairs. The fruits are narrow, beaked pods, usually 2–5 cm (1–2 in.) long, each containing 4–10 small "peas."

OCCURRENCE A widespread weed of gardens, waste places, meadows, and hillsides throughout North America; sometimes cultivated for animal feed as a hay crop and for silage.

TOXICITY Various species of Vetch, and particularly Common Vetch, have been reported from time to time to produce disease and loss of life in humans and livestock. The main toxic components are cyanogenic glycosides, which release hydrocyanic acid in the presence of enzymes in the plant or in the digestive tract. Different species, varieties, populations, and life cycle stages of Vetches exhibit differing concentrations of the glycosides and the enzymes that break them down. The seeds of Common Vetch have been shown to contain lethal levels of cyanogenic glycosides in some populations, and one

Fig. 57. Common Vetch, fruiting pods. R. & N. TURNER

subspecies, Narrow-leaved Vetch (*V. sativa* ssp. *angustifolia*), formerly treated as a separate species, is known also to contain a toxic amino acid similar to those causing toxicity in Wild Peas (*Lathyrus* spp.). Symptoms of Vetch poisoning are typical of cyanide poisoning: rapid breathing followed by difficulty in breathing, excitement and restlessness, and in severe cases, prostration, convulsions, coma, and death.

TREATMENT Induce vomiting, or perform gastric lavage; if large quantities have been eaten, or if serious symptoms occur, treat for cyanide poisoning (see p. 12).

NOTES Because Vetch pods and seeds resemble those of garden Peas, there is a danger of children eating them in play. There are many species of Vetch occurring in North America, some indigenous and some introduced and naturalized, or grown as forage. Another species implicated in many instances of fatal cattle poisoning is Hairy Vetch (*Vicia villosa*). All Vetches should be considered potentially toxic, and should be treated with caution, until more is known about individual types. Broad Bean is another species of Vetch (*Vicia faba*). It is not normally poisonous, but can cause severe hemolytic anemia in some individuals.

WILD FLOWERING PLANTS

Baneberry, or Dolls-eyes
(Buttercup Family)

Actaea rubra and *A. pachypoda*
(Ranunculaceae)

Fig. 58. Baneberry (*Actaea rubra*), red and white color forms. R. & N. TURNER

QUICK CHECK Bushy, herbaceous plants with compound leaves, small, whitish flowers in dense, elongated clusters, and bright red or white berries. All parts of the plants toxic and potentially fatal.

DESCRIPTION Herbaceous plants up to 1 m (3 ft) tall, growing from thick rhizomes. The leaves are large, spreading, and coarsely twice or 3 times divided into 6 or more pointed, sharply toothed leaflets. The flowers are small and whitish, borne in dense, long-stalked terminal clusters, or racemes. The berries are bright red or white, and very showy. *Actaea rubra* usually has red berries, but a white-fruited form sometimes occurs. The berry stalks of *A. rubra* are very slender. *A. pachypoda* is usually white-fruited, rarely red-fruited, and its berry stalks are stout, or swollen.

OCCURRENCE Western Baneberry, *Actaea rubra*, is found from Alaska to central California and west to the central United States. *A. pachypoda* occurs from Canada south to Georgia and Louisiana, and east to the northern Rockies. Both species grow in moist, rich woods and along creeks.

TOXICITY All parts of the plants, especially the roots and berries, are toxic. Eating only six berries can produce severe symptoms. Baneberries, like other plants in the buttercup family, contain protoanemonin, but the toxicity of the plants is apparently mainly due to an as yet undetermined toxin, probably an essential oil or poisonous glycoside. Symptoms, lasting up to about three hours, include acute stomach cramps, dizziness, vomiting, increased pulse, delirium, circulatory failure, and headache. No deaths have been reported for the North American species, but there are several documented cases of fatal poisoning in

Fig. 59. Baneberry (*Actaea pachypoda*), fruit.
AUDREY B. BURNAND, COURTESY ROYAL BRITISH COLUMBIA MUSEUM, VICTORIA

children from eating a black-fruited European species of Baneberry (*A. spicata*).

TREATMENT Induce vomiting, or perform gastic lavage; follow with activated charcoal and a saline cathartic; treat for symptoms; general supportive therapy.

NOTES The plant is used medicinally by some Native Indian groups, usually in the form of a decoction which is drunk. The Lillooet Indians of British Columbia used Baneberry (*Actaea rubra*) as a general tonic, but their name for the plant translates as "it-makes-you-sick," and they were very aware of its poisonous qualities.

Bleedingheart, Wild, Dutchman's-Breeches, and Squirrel-Corn
(Fumatory Family)

Dicentra spp.
(Fumariaceae)

Fig. 60. Western Bleedingheart (*Dicentra formosa*).
R. & N. TURNER

QUICK CHECK Delicate, herbaceous, early spring-flowering plants with finely divided, fern-like leaves and hanging, clustered, two-sided flowers. Occasional cause of human poisoning but no reported fatalities.

Fig. 61. Dutchman's-Breeches
(*Dicentra cucullaria*).
MARY W. FERGUSON

DESCRIPTION Herbaceous plants up to 60 cm (2 ft) high, growing from a fleshy, sometimes branching rootstock. The leaves, growing from the rootstock, are long-stalked, broadly triangular, and finely dissected, giving the plants a delicate, lacy appearance. The showy flowers are loosely clustered, often hanging, each with 4 petals fused into 2 flattened pairs, rounded at the tops, or prolonged backwards into 2 prominent spurs. About a dozen species occur in North America, most potentially poisonous. Most prominent are: Dutchman's-Breeches (*Dicentra cucullaria*), with bulbous rootstock, and white or pinkish, yellow-tipped flowers that resemble baggy pants hanging upside-down; Squirrel-Corn (*D. canadensis*), with tuber-producing rhizomes and greenish-white or pinkish heart-shaped flowers; and Wild Bleedingheart (*D. eximia*) and Western Bleedingheart (*D. formosa*), both with branching rhizomes and rose-pink to purplish heart-shaped flowers. The garden Bleedingheart (*D. spectabilis*) is also potentially poisonous.

OCCURRENCE The Bleedinghearts and their relatives are found in moist, rich woodlands and meadows throughout North America, and are often grown in gardens. Dutchman's-Breeches is found in eastern North America from Nova Scotia to Missouri and Alabama. Squirrel-Corn grows in the East from Nova Scotia to North Carolina and Tennessee. Wild Bleedingheart is found in the southern Appalachians, and Western Bleedingheart in the West from British Columbia to California.

TOXICITY Various species of Bleedinghearts have been shown to contain isoquinoline alkaloids, including aporphine, protoberberine, and protopine. They are known to be toxic to both humans and animals. Symptoms of poisoning include: trembling, agitation, heavy salivation, vomiting, diarrhea, convulsions, tenseness of muscles, difficult breathing, and prostration. Fortunately, human poisoning is rare, and recovery is usually rapid and complete.

Although not reported to have caused human deaths, the Bleedinghearts are potentially harmful to children, since they are showy and attractive and are often found around meadows, in gardens, or as potted ornamentals. Contact with the plants can cause an allergic skin reaction in some individuals.

TREATMENT Induce vomiting, or perform gastric lavage; follow with activated charcoal and a saline cathartic; treat for symptoms; tranquilizers may be required.

NOTES Members of the genus *Corydalis,* related to *Dicentra,* also contain toxic isoquinoline alkaloids. Several *Corydalis* species are native to North America, and they are also sometimes grown as garden flowers. Although not implicated in human poisoning, they have caused poisoning and death in animals. Livestock loss from Bleeding-hearts and their relatives usually occurs on early spring woodland pasture. Cattle find the plants distasteful and eat them only when other food is scarce. Heavy trampling of the ground when soft may expose the fleshy rootstocks which may then be eaten with harmful results.

Bloodroot, or Red Puccoon
(Poppy Family)

Sanguinaria canadensis
(Papaveraceae)

Fig. 62. Bloodroot (*Sanguinaria canadensis*).
MARY W. FERGUSON

QUICK CHECK White-flowered herbaceous plant with large, palmately lobed leaves; rhizome, leaves, and stems exude a bright red-orange sap when cut. Highly toxic; potential poisoning from herbal medicine preparations.

DESCRIPTION Herbaceous perennial growing from a stout, knotted rhizome. The leaves are single, large and long-stalked, and palmately lobed with 3–9 rounded lobes. The flowers are white and conspicuous, 2–5 cm (1–2 in.) across, growing singly on a stalk up to 15 cm (6 in.) high. They bear 8–16 petals, with 4 usually longer than the others. The fruit is an elongated capsule up to 5 cm (2 in.) long. The rhizome, leaves, and stems all exude a red-orange sap, or latex, when bruised.

OCCURRENCE Common, well known spring wildflower of rich woods of eastern and central North America from southern Canada to Florida and Texas.

TOXICITY The red-colored latex found throughout this plant contains several alkaloids, including sanguinarine, chelerythrine, protopine, and homochelidonine, as well as resins. These physiologically active compounds can, if consumed, cause vomiting, diarrhea, fainting, shock, coma, and potentially death. They can also cause fluid retention, or edema, and glaucoma. However, there are apparently no known cases of human or livestock poisoning by Bloodroot under natural conditions.

TREATMENT Induce vomiting, or perform gastric lavage; follow with activated charcoal and a saline cathartic; treat for symptoms.

NOTES The red latex was formerly extracted from the rootstock and used as a medicinal drug. It is now also used in research to induce glaucoma in laboratory animals. The same alkaloids contained in Bloodroot latex are present in other members of the poppy family (see p. 206).

Buttercups, or Crowfoots
(Buttercup Family)

Ranunculus spp.
(Ranunculaceae)

Fig. 63. Creeping Buttercup (*Ranunculus repens*).
R. & N. TURNER

QUICK CHECK Low to moderately tall, herbaceous plants with variously lobed or entire leaves, many species with attractive, yellow flowers. Fresh plants contain an irritant oil which can cause painful blistering of the skin and irritation of the mouth and digestive tract; human fatalities unknown.

DESCRIPTION The Buttercups are a very large group of annual or perennial herbs. The stem leaves are alternate, palmately veined, entire, and lobed or finely divided. Basal leaves are often present, numerous, and variable in shape. The flowers are usually yellow, sometimes cream-colored, and are borne singly or in loose terminal clusters, each with 5 (or sometimes more) petals. The fruits consist of a group of small, 1-seeded achenes on a rounded receptacle.

OCCURRENCE Buttercups are usually spring-flowering, and are found throughout North America in a variety of habitats, from open fields, clearings, and gardens, to moist woods. Some are aquatic; many are common weeds.

TOXICITY The Buttercups, and other members of the buttercup family such as Anemones (*Anemone* spp. and *Pulsatilla* spp.) and Marsh Marigolds (*Caltha* spp.), contain varying quantities of an acrid, blister-causing juice which yields a highly irritant yellow oil, protoanemonin. This substance is produced in the plant through the enzymatic breakdown of a glycoside, ranunculin. Although it is present throughout the tissues of fresh plants, especially during flowering, protoanemonin is unstable, and changes, when the plants are dried, into an innocuous form, anemonin. Protoanemonin can cause severe irritation and blistering of the skin, or, if consumed, severe gastrointestinal irritation.

Buttercups and their relatives, especially those with showy flowers, are potentially poisonous to humans, and some species have been known to poison children, causing burning of the mouth, abdominal pain, and diarrhea. However, these plants are mainly known for their toxicity to grazing animals. In severe cases they can be fatal, but since they are strongly distasteful, they are seldom eaten in quantity.

TREATMENT If raw Buttercup has been taken internally, induce vomiting; give demulcents such as milk to sooth the digestive tract; treat for symptoms; gastric lavage is seldom required.

NOTES Buttercup species vary in the quantities of protoanemonin they contain. Known poisonous species include Tall Field Buttercup (*Ranunculus acris*), Bulbous Buttercup (*R. bulbosus*), Small-flowered Buttercup (*R. abortivus*), Creeping Buttercup (*R. repens*), and Cursed Crowfoot (*R. sceleratus*).

Many North American Indian groups traditionally used Buttercups and their relatives as counter-irritant medicines, applied externally to draw out pain and infection from underlying tissues. They were aware of their potentially poisonous properties, however, and did not usually use them internally.

**This book cannot replace the advice and assistance
of qualified medical personnel.
In all cases of suspected poisoning by plants, or any other substance,
immediate qualified medical advice and assistance should be sought.**

Cocklebur
(Aster, or Composite Family)

Xanthium strumarium and
related species
(Asteraceae, or Compositae)

Fig. 64. Cocklebur (*Xanthium strumarium*).
WALTER H. LEWIS

QUICK CHECK Coarse, weedy annual with palmately lobed leaves and thick, prickly burs along the upper stem of mature plants. Seeds and seedlings toxic to animals and potentially fatal; no human poisonings reported.

DESCRIPTION Coarse, herbaceous, annual weed up to 1 m (3 ft) or more high, with erect, stout, branching stems. The leaves are alternate, stalked, and pointed, with toothed or irregularly lobed margins. The flowerheads lack colored rays, and are separated into small, many-flowered male (pollen-producing) heads near the tops of the stems and 2-flowered female (seed-producing) heads clustered at the leaf axils below. The fruits are conspicuous, elongated, 2-chambered burs with dense, hooked prickles, which attach easily to clothing and animal fur. This is a highly variable species, and includes several varieties recognized by some as separate species: *Xanthium pennsylvanicum*, *X. chinense,* and *X. speciosum,* for example. Another well-known species is *X. spinosum,* with narrower, often deeply lobed leaves.

OCCURRENCE Introduced from Europe, Cocklebur is a widespread weed of fields, waste places, shorelines, and flood plains; it is found throughout North America. The seeds sprout readily when present in soil that has recently been under water.

TOXICITY The main toxic agent of this plant, concentrated in the seeds and seedlings, is called carboxyatractyloside, a highly toxic glycoside. Widely known as a livestock poison, it produces symptoms including loss of appetite, digestive tract inflammation, excitement, weakness, loss of coordination, prostration, and in severe cases, convulsions and death. Human poisoning from Cocklebur has not been reported, but children should be kept away from it. It is also known to cause allergic skin irritation on contact.

TREATMENT Induce vomiting, or perform gastric lavage; general supportive treatment for symptoms; administer fatty substances, such as milk or lard.

NOTES Another plant in the same family, Burdock (*Arctium* spp.), may be confused with Cocklebur, but it is usually larger, and its burs are more spherical. The burs of both Burdock and Cocklebur can cause mechanical injury, and Burdock causes dermatitis in some people, but it is not considered to be poisonous to eat; in fact, its taproot and peeled stems are a well-known food in Japan and elsewhere.

Death Camas, or Black Snakeroot
(Lily Family)

Zigadenus venenosus and other *Zigadenus* species (also spelled *Zygadenus*)
(Liliaceae)

Fig. 65. Death Camas (*Zigadenus venenosus*).
R. & N. TURNER

QUICK CHECK Perennial with bulb, basal and grass-like leaves, and showy, cream-colored flowers in a dense, terminal cluster. Entire plant, especially onion-like bulbs, highly toxic and sometimes fatal for humans; extremely toxic to grazing livestock.

DESCRIPTION There are about 15 species of *Zigadenus* in North America, known generally as Death Camas, or Black Snakeroot. They vary in toxicity, but all should be considered potentially poisonous. One of the commonest and best known is *Z. venenosus*. It is a perennial lily usually 30–60 cm (1–2 ft) high, growing from a bulb which resembles a small onion, but lacks any onion odor. The leaves are smooth, grass-like, and V-shaped in cross-section, mostly growing from the base. The 6-petalled flowers are creamy white, and crowded together in a dense, terminal cluster. The fruits are cylindrical, 3-parted capsules. The stalk tends to elongate at the fruiting stage, so that the capsules are further spread apart than the flowers. Other species have greenish, white, yellow, or pink flowers, in some cases more loosely clustered; some grow from rhizomes.

OCCURRENCE *Zigadenus venenosus* is common in moist, grassy meadows in western North America, from southern British Columbia to California, and east to Utah and Nevada. Flowers in May and June. Other species occur throughout North America from Alaska to Florida.

TOXICITY Death Camas species contain several toxic alkaloids, similar to those found in Indian Hellebore (*Veratrum* spp.). These include zygacine, zygadenine, iso- and neogermidine, and protoveratridine. The entire plants, including the bulbs and flowers, are poisonous.

Human poisonings from Death Camas are rare, but sometimes occur when people mistake the bulbs for those of edible species such as Blue Camas (*Camassia* spp.) or Wild Onions (*Allium* spp.). Death Camas species have been responsible for losses of large numbers of livestock, especially sheep and cattle.

Humans poisoned by Death Camas may experience excessive watering of the mouth, burning followed by numbness of the lips and mouth, thirst, headache, dizziness, nausea, stomach pain, persistent vomiting, diarrhea, muscular weakness, confusion, slow and irregular heartbeat, low blood pressure, subnormal temperature, and in severe cases, difficulty in breathing, convulsions, coma, and death. Symptoms of poisoning may appear from 1–8 hours after eating Death Camas, depending on the species. Symptoms in animals are similar. Recovery usually occurs within 24 hours.

TREATMENT Induce vomiting, or perform gastric lavage; follow with activated charcoal and saline cathartic; general supportive treatment for symptoms; maintain fluid and electrolyte balance; monitor breathing, heart rate, blood pressure; subcutaneous application of atropine, repeated as needed, will alleviate slow heart rate; for persistent low blood pressure, ephedrine or dopamine may be given; control convulsions with i.v. diazepam.

NOTES Most Native people who eat bulbs such as Blue Camas (*Camassia* spp.) and Mariposa Lily (*Calochortus* spp.) as part of their traditional diet, are well aware of Death Camas, and are careful to avoid the bulbs in their harvesting. Some call them "poison onions." Still, there are a number of stories of accidental poisonings among Native peoples. Aside from *Zigadenus venenosus,* the most toxic species of Death Camas include *Z. gramineus, Z. paniculatus, Z. nuttallii,* and *Z. densus.* Another bulb plant in the lily family, Mountain Bells (*Stenanthium occidentale*), with purplish flowers, is believed by some Native peoples to be very poisonous as well.

Fig. 66. Death Camas bulb and seed stalk. IAN G. FORBES

Grasses, Wild
(Poa, or Grass Family)

Various species
(Poaceae, or Gramineae)

Fig. 67. Perennial Rye Grass (*Lolium perenne*).
R. & N. TURNER

QUICK CHECK Grasses are variable in size and flowerhead shape, but all have jointed stems, long, slender, parallel-veined leaves, chafey, petalless flowers in clusters, and seed-like fruits known as grains. Some species have potentially toxic grains; all grains can be infested with toxic molds or other fungi; some grass fruits can cause choking if swallowed.

DESCRIPTION Grasses are well known and important economic plants having jointed stems, slender, sheathing, parallel-veined leaves, and usually small, inconspicuous flowers borne in bracted spikelets that are variously clustered into loose or compact heads. Although they include some of the most important edible plants of the world—the cereal grains (Wheat, Maize, Rice, Oats, Rye, and Barley)—some grasses are potentially harmful. It is not possible here to include descriptions of individual types, but some are mentioned as potentially dangerous.

OCCURRENCE Grasses are common and widespread in North America, growing in almost every environment, and recognized by most people as belonging to a distinct group.

TOXICITY Several types of poisoning have been attributed to grasses, mostly related to their widespread use as forage and hay crops for animals. The deadly fungus, Ergot, which commonly infects cereal grains, is discussed elsewhere (p. 64). Other fungi, including molds such as species of *Penicillium*, are also known to render toxic grass species that are otherwise edible.

Fungi have been implicated in the toxicity of one well known poisonous grass genus, *Lolium*, which includes an annual species,

Fig. 68. Brome Grass, or Cheat Grass (*Bromus tectorum*), showing sharp-awned fruiting heads.
R. & N. TURNER

Darnel (*L. temulentum*), and Perennial Rye Grass (*L. perenne*). Records go back to ancient times of people being poisoned by eating flour or bread contaminated with Darnel grains, and occasional livestock poisonings from this grass have also been reported. Perennial Rye Grass is known to cause a nervous disorder, called Ryegrass staggers, in grazing animals.

Some grasses, such as Velvet Grass (*Holcus lanatus*) and Manna Grass (*Glyceria* spp.), contain cyanogenic (cyanide-producing) compounds, but these are present in low concentrations and are usually harmless.

Other grass species, such as Foxtail Barley, or Squirreltail (*Hordeum jubatum*), Brome Grass (*Bromus* spp.), Porcupine Grass, or Needle-and-Thread (*Stipa* spp.), and Bristle Grass, or Foxtail Grass (*Setaria* spp.), have long, sharp, wiry bristles or awns associated with their fruiting heads. These awned fruits are often barbed, and can easily become lodged in the ears, noses, and throats of animals and people. Once they penetrate, they can be difficult to remove. In the summer and fall dogs often must be taken to a veterinarian for removal of grass fruits from deep inside their ears. Berry pickers should be careful to remove any awned grass fruits from their harvest, because they can cause choking if swallowed.

Cutting sharply awned grasses before they go to seed, and not allowing dogs and other animals into areas where these grasses are present during the danger period (mid-summer to fall) can reduce the hazard of mechanical injury from sharp grass fruits. Close inspection of fur and ears after animals have been in a danger area is also recommended.

Various other toxicological problems associated with grasses are discussed by Kingsbury (1964), Cooper and Johnson (1984), and Fuller and McClintock (1986), but these are usually seen only in exceptional combinations of circumstances.

TREATMENT General supportive measures for symptoms.

NOTES One toxic syndrome related to grass pasturage is known variously as *grass tetany, grass staggers, wheat pasture poisoning, protein poisoning,* and *hypomagnesemia.* It has caused livestock loss, especially of cattle, in various parts of North America, and is associated with pasturing of cattle on lush young growth of pasture or range grass, or wheat forage, in early spring or during mild winters in some areas. The exact cause of this disorder is not known but is believed to be related to an inability of the animals to absorb magnesium, resulting in a deficiency of this element.

Large quantities of domesticated cereal grains—Wheat, Oats, Barley, Rye, Sorghum, and Maize, or Corn—are used as livestock feed in North America. They sometimes cause digestive disorders and other problems, but, with the exception of Rye (which contains some growth-depressing substances and one that produces rickets in chicks), none produces any toxic substances. Any problems with them are caused by fungal contamination due to poor storage, or to poor feeding practices. The best means of preventing grass-related poisoning is to provide animals with good, properly tended pasturage and properly stored feed.

Indian Hellebore, or False Hellebore
(Lily Family)

Veratrum viride (syn.
   V. eschscholtzii)
(Liliaceae)

Fig. 69. Indian Hellebore (*Veratrum viride*), young shoots. R. & N. TURNER

QUICK CHECK Tall, erect herbaceous perennial with large, alternate, parallel-veined, distinctly pleated leaves and greenish or whitish flowers in dense, terminal clusters. All parts of plant highly toxic, potentially fatal; occasional human poisoning, most from misuse of medicinal preparations.

DESCRIPTION Leafy, unbranched, herbaceous perennial about 1–2.5 m (3–8 ft) high, growing from a thick rootstock. The leaves are oval, pointed, and densely hairy beneath, up to 30 cm (1 ft) long and 15 cm (6 in.) wide near the base of the plant, becoming smaller and narrower towards the top. The 6-parted flowers, up to 2 cm (0.8 in.)

Fig. 70. Indian Hellebore, mature
plant. R. & N. TURNER

Fig. 71. Indian Hellebore, closeup
of flowers.   IAN G. FORBES

across, are hairy, and greenish-white or greenish-yellow, often with a
dark central blotch. They grow in dense, branching, terminal clusters
30 cm (1 ft) or more long, the branches spreading and somewhat
drooping. The fruits are 3-parted capsules, straw-colored to dark
brown. Other *Veratrum* species, including *V. californicum,* with whitish
flowers and erect flower clusters, and *V. parviflorum,* with narrower
upper leaves and hairless flowers, are also known to be poisonous.
European and Asian species are also toxic.

OCCURRENCE Rich, moist woods and mountain meadows along streams
and wet areas, from Alaska to northern Oregon and eastward to the
Atlantic coast. *Veratrum californicum* occurs in mountain meadows and
valleys of the western states from the Pacific Coast to the Rocky
Mountains, and *V. parviflorum* in the mountains of the eastern states
from West Virginia to Georgia.

TOXICITY Indian Hellebore and its relatives contain numerous, complex
alkaloids, including germidine, germitrine, veratridine, veratrosine,
and veratramine. All parts of the plant are toxic, but the highest
concentration of toxins is said to be in the inner rhizome.

Most instances of *Veratrum* poisoning in humans have been as a
result of misuse of medicinal preparations containing the plant. There
are scattered reports of fatalities in humans and various types of
livestock from eating the plant itself. Symptoms of poisoning include:
burning sensation of mouth and throat, and pain in upper abdomen,
followed by watering of the mouth, vomiting, diarrhea, sweating,
blurred vision, hallucinations, headache, general paralysis, and
spasms. In severe cases, shallow breathing, slow or irregular pulse,

lower temperature, convulsions, and death may occur. The combined symptoms have been characterized as similar to those of a heart attack. Symptoms usually disappear within 24 hours.

The greatest danger of livestock poisoning by Indian Hellebore is in early spring, when it produces young, succulent shoots, and when other forage is scarce.

TREATMENT Induce vomiting, or perform gastric lavage; follow with activated charcoal and a saline cathartic; low blood pressure is counteracted by fluid replacement and, if necessary, dopamine or ephedrine. Monitor heart beat and blood pressure; maintain fluid and electrolyte balance. Atropine is recommended for slow heart beat. Breathing assistance and oxygen may be required. Recovery usually occurs within 15 hours.

NOTES *Veratrum* has been used medicinally for treating high blood pressure; small doses are used to reduce blood pressure with no noticeable effect on respiratory or cardiac rate. Unfortunately, the complexity and relative instability of the alkaloids in this plant make the drug difficult to standardize.

Native people in many areas of the continent, particularly the Pacific Northwest, were well aware of the toxic and medicinal properties of Indian Hellebore. Its rootstock was widely used as an external medicine and local anaesthetic for innumerable ailments, including arthritis and bruises. It was also sometimes taken, with great caution and in very diluted form, as a cleansing purgative during certain rites of hunters, shamans, and others seeking special powers. However, there are a number of reports of accidental poisoning by this plant. Some British Columbia Natives have stressed that the only effective antidote for poisoning from Indian Hellebore is eating large amounts of salmon oil. This treatment was also used for Death Camas (p. 106) and Water Hemlock (p. 141) poisoning.

Veratrum californicum is the cause of a usually fatal type of birth deformity in lambs, known as "monkey-face," traced to ewes feeding on this plant during their early pregnancy in late summer.

Prolonged numbness of the mouth and nausea from drinking water in which Indian Hellebore was growing has been reported.

Indian Hemp
(Dogbane Family)

Apocynum cannabinum
(Apocynaceae)

Fig. 72. Indian Hemp (*Apocynum cannabinum*).
R. & N. TURNER

QUICK CHECK Tall, bushy herb of roadsides and clearings, with opposite, smooth-edged leaves, small, clustered whitish flowers, long thin seed pods, and milky sap. Potentially highly toxic; possible human poisoning from herbal medicine preparations.

DESCRIPTION Indian Hemp is an erect, bushy herbaceous perennial 1–1.5 m (3–5 ft) tall, growing from spreading rhizomes. The leaves are simple, opposite, smooth-edged, and oval to elliptical. The flowers are small, whitish, and urn- or bell-shaped, borne in clusters toward the tops of the stems, and the fruits are long, narrow, paired pods containing numerous seeds attached to long, silky, milkweed-like hairs. The stems are very fibrous, and exude a milky juice when broken. A smaller, pink-flowered species, Spreading Dogbane (*A. androsaemifolium*), has similar characteristics.

OCCURRENCE Common summer-flowering weedy plant of roadsides, pastures, and clearings, widely distributed in North America. Spreading Dogbane is also common in many parts of the continent.

TOXICITY Indian Hemp and its relatives contain several resins and glycosides, some cardioactive, including cymarin, apocannoside, and cyanocannoside. Cases of poisoning to humans have not been reported, but their toxicity to animals is known, mainly through laboratory testing. The plants are distasteful to animals and not usually consumed. One glycoside, apocynamarin, is known to increase blood pressure, and some of the resins caused gastric disturbance and death in a dog. The milky sap may cause dermatitis.

TREATMENT In case of suspected poisoning, induce vomiting, or perform gastric lavage; treat for symptoms.

NOTES Indian Hemp and Spreading Dogbane were officially recognized in the nineteenth century as diaphoretics and expectorants, with emetic, cathartic, and diuretic properties. They were also used by Native people as medicines for kidney ailments and other afflictions. However, their pharmacological effects are probably due to their toxic cardiotonic activity, and they should be used only with extreme caution.

Indian Tobacco
(Lobelia Family)

Lobelia inflata
(Lobeliaceae)

Fig. 73. Indian Tobacco (*Lobelia inflata*). FRANK FISH

QUICK CHECK Erect, branching annual weed, with inconspicuous blue flowers, and inflated seed pods. Highly toxic; potentially fatal. Most human poisonings from misuse of herbal medicine application.

DESCRIPTION A branching annual up to 1 m (3 ft) tall, with simple, alternate, toothed leaves tending to run down the stem. The flowers, clustered at the ends of the stem and branches, are small, pale blue to white, tubular, and bilaterally symmetrical, with two lobes above and three below, the tube split nearly to the base along the top. The fruit is a capsule, conspicuously inflated and bladderlike. There are several other species of *Lobelia* in North America, including the showy, red-flowered Cardinal Flower (*Lobelia cardinalis*), and Blue Cardinal Flower (*L. siphilitica*) (see under Garden Flowers). All Lobelias should be considered toxic.

OCCURRENCE Indian Tobacco is a common herbaceous weed of fields, woods, and waste places throughout eastern North America. Red and Blue Cardinal Flowers occur along water courses and in damp soil in southeastern Canada and the eastern United States, and are frequently cultivated as ornamentals, as are several other species of *Lobelia*.

TOXICITY Several species of *Lobelia*, including Indian Tobacco and the Cardinal Flowers, are known to be toxic to both humans and livestock, due to the presence of a variety of pyridine alkaloids, especially lobelamine and lobeline. Over 14 of these alkaloids have been isolated from Indian Tobacco alone. Similar in structure to nicotine, they produce symptoms in humans including nausea, progressive vomiting, sweating, pain, tremors, weakness, paralysis, depressed temperature, rapid but weak pulse, and, in serious cases, convulsions, coma, and death from respiratory failure. Most human poisoning results from inappropriate use of Lobelias or their extracts as medicinal preparations.

Livestock poisoning is not common, but does occur occasionally, with symptoms appearing after about three days of eating Lobelia. The entire plants, including extracts of the leaves or fruits, are potentially poisonous. *Lobelia* species can also cause skin irritation in humans.

TREATMENT Do not induce vomiting or perform gastric lavage if vomiting has occurred or if seizures are imminent; i.v. diazepam may be used to sedate patient, then administer activated charcoal and a saline cathartic (repeat charcoal and cathartic every 6 hours for 24 hours); breathing assistance and oxygen may be required; monitor heart, arterial blood gases; anti-convulsive therapy, with i.v. diazepam or phenobarbital, and general supportive therapy for symptoms are suggested.

NOTES The leaves of Indian Tobacco were dried and smoked by Native peoples of eastern North America. The plant was also used in folk medicine, and pharmacologically as an emetic, expectorant, and respiratory stimulant. Additionally, it has been employed as a deterrent to smoking tobacco.

Jack-in-the-Pulpit, or Indian Turnip
(Arum Family)

Arisaema atrorubens,
   A. stewardsonii, and
   A. triphyllum
(Araceae)

Fig. 74. Jack-in-the-Pulpit (*Arisaema stewardsonii*).
FRANK FISH

QUICK CHECK Distinctive herbaceous perennials with 3-parted leaves and brownish or greenish "flower" consisting of a central cone-like spike encased within a hood-like spathe. Corms and entire plant contain irritants that cause intense burning and inflammation to mouth and throat; not usually fatal.

DESCRIPTION Herbaceous perennials growing from a swollen corm. The paired, long-stalked leaves are compound, and usually 3-parted, with pointed, smooth-edged leaflets. The flowers are tiny, and crowded together in a short, blunt cluster, or spadix, surrounded by an ensheathing spathe, which is purplish, brown or green and striped with brown, and shaped like a pulpit with a long, oval-shaped hood extending over the spadix. The fruits are scarlet berries borne in a dense, egg-shaped cluster. The three species mentioned are closely related and very similar. *Arisaema stewardsonii* has a conspicuously whitish-ridged spathe; the other two are distinguished on the basis of the hood, which is broader in *A. atrorubens* that in *A. triphyllum,* and the fruiting head, which is less than about 2 cm (1 in.) long in *A. triphyllum* and more than 3 cm (over 1 in.) in the other two. Another related species, called Green Dragon (*A. dracontium*), is similar, but has a 7–13 parted leaf, with a pointed spadix extending well beyond the summit of the greenish spathe. All four species have similar irritant properties.

OCCURRENCE Jack-in-the-Pulpit and related species are found in rich woods, thickets, swamps, and bogs throughout much of eastern North America. *Arisaema triphyllum* occurs only in the eastern United States; the other three species also occur in southeastern Canada.

TOXICITY The entire plants, including the corms, contain numerous microscopic bundles of needle-like crystals of calcium oxalate. When the plants or corms are eaten fresh, these pierce the tender tissues of the mouth, tongue, and throat, and cause intense burning and inflammation. In serious cases, choking may result from swelling of the throat. Irritation is also attributed to several proteolytic enzymes, which trigger the release of kinins and histamines by the body. Salivation, nausea, vomiting, and diarrhea may occur, and, very rarely, irregular heartbeat, dilation of the pupils, fits, coma, and potentially, death. Because of the immediate reaction, rarely more than the first mouthful is consumed, and hence the plant is seldom life-threatening. The leaves and roots can cause skin irritation on contact.

Fig. 75. Jack-in-the-Pulpit (*Arisaema atrorubens*), fruit.
MARY W. FERGUSON

Fig. 76. Eastern Skunk-Cabbage (*Symplocarpus foetidus*). MARY W. FERGUSON

Fig. 77. Western Skunk-Cabbage (*Lysichitum americanum*). R. & N. TURNER

TREATMENT Symptoms rarely require intensive treatment; the irritation and swelling recede after a few hours. Treat for symptoms; give demulcents such as milk; cool liquids or ice held in the mouth may bring some relief; antihistamines or epinephrine may be given. In the event of choking, artificial respiration may be required. In case of skin irritation from touching the plant, give antihistamines, and/or apply a recognized brand of topical ointment or cream.

NOTES Many species in the arum family have similar irritating crystals in their leaves and roots, including a number grown as ornamental house plants. Three other wild species of this family, Eastern Skunk-Cabbage (*Symplocarpus foetidus*), Western Skunk-Cabbage (*Lysichitum americanum*), and Water Arum (*Calla palustris*) are similarly toxic. All are herbaceous perennials of swampy ground. The skunk-cabbages have large, fleshy leaves giving off a pungent, skunk-like odor when bruised. Their numerous small flowers form a dense, club-like head surrounded by a large sheath, or spathe. Blooming time is in early spring. In Eastern Skunk-Cabbage, the spathe is purplish or brownish mottled; in Western Skunk-Cabbage it is bright yellow and showy. Water Arum has long-stalked heart-shaped leaves, with a smaller, inconspicuous greenish-white flower spathe, and red berries borne in dense clusters.

Despite the irritant properties of Jack-in-the-Pulpit corms they were used as a source of flour by Native peoples and early pioneers, after being thoroughly dried and pulverized to break up the calcium oxalate crystals. Boiling alone does not dispel them. The rootstocks and young leaves of the Skunk-Cabbages were also sometimes eaten after cooking, but since some residual poison remains, their use is not recommended.

Jimsonweed, Downy Thornapple, and Relatives
(Nightshade Family)

Datura stramonium, D. meteloides,
   and related species
(Solanaceae)

Fig. 78. Downy Thornapple (*Datura meteloides*).
R. & N. TURNER

QUICK CHECK Large, weedy annuals or perennials with big, irregularly toothed leaves, white or purplish tinged funnel-shaped flowers, and prickly, spherical or egg-shaped fruiting capsules. Entire plants highly toxic. Sometimes misused in herbal medicine or as hallucinogens; potentially fatal.

DESCRIPTION Jimsonweed (*D. stramonium*) is a large, branching, annual weed 1–1.5 m (3–5 ft) tall, ill-smelling, and generally lacking hairs. The leaves, 8–20 cm (3–8 in.) long, are alternate, oval, and irregularly toothed or lobed. The showy, funnel-shaped flowers are erect, up to 10 cm (4 in.) long, and white (or purplish in one variety), with a 5-pointed rim. The fruit is an ovoid, spiny capsule, splitting regularly into 4 parts, and bearing numerous wrinkled, black seeds. Downy Thornapple (*D. meteloides*) is a bushy, branching perennial, up to 1 m (3 ft) tall and 2 m (6 ft) across, with a grayish green coloring from a dense covering of hairs. The leaves and flowers are similar to those of Jimsonweed, but generally larger. The spherical capsules are nodding or reflexed, with short, hairy spines, splitting open irregularly when ripe.

There are a number of related species, also toxic, including: Desert Thornapple (*D. discolor*), of the Colorado Desert area of California; Metel, or Devil's Trumpet (*Datura metel*), from India but introduced and cultivated widely; and Angel's Trumpet (*D. suaveolens*), an ornamental shrub or small tree cultivated in warmer regions, with large, trumpet-like flowers.

OCCURRENCE Jimsonweed is widely naturalized around the world, and is presumed native to the eastern United States. It is now a common and widely distributed weed of fields, roadsides, and waste places throughout much of the United States and southern Canada. Downy Thornapple is native to southwestern United States and Mexico.

TOXICITY All parts of *Datura* plants, especially the leaves and seeds, contain several toxic indole alkaloids, mainly hyoscyamine (an isomer of atropine which blocks the parasympathetic nervous system) and

Fig. 79. Downy Thornapple, spiny fruit capsules. R. & N. TURNER

hyoscine, or scopolamine, which is hallucinogenic. Symptoms of poisoning, similar to those of Belladonna (*Atropa belladonna*), may appear from a few minutes to several hours after ingesting the plant. They include intense thirst, dilation of pupils, blurred vision, flushing and dryness of the skin, headache, nausea, rapid but weak pulse, high temperature (occasionally), high blood pressure, urinary retention, hallucinations, delirium, incoherence, fever, convulsions, coma, and death. Intense symptoms may abate after 12–48 hours, but disturbance of vision may last up to two weeks. There have been many of cases of human poisoning: of children, from sucking the flower nectar or eating the seeds; of those drinking a "tea" from the leaves as a medicinal preparation for relief of asthma and other ailments; or of people deliberately seeking a hallucinogenic experience from the plant. Eating less than 4–5 gm (1/5 oz.) of the seeds or leaves can be fatal to a child.

Jimsonweed leaves and flowers can cause skin irritation when handled.

Jimsonweed, Downy Thornapple, and their relatives are also toxic to all classes of livestock, and poisoning has occurred when the plants contaminate hay or the seeds become inadvertently mixed in with grain for poultry. Symptoms in livestock are similar to those in humans. Atropine can often be detected in the stomach contents and body tissues for some time after the death of a poisoned animal.

TREATMENT Induce vomiting, or perform gastric lavage; follow with activated charcoal and a saline cathartic; monitor breathing, heart, and blood pressure; Breathing assistance and oxygen may be required; catheterizing may be necessary; propranolol is suggested for irregular heart beat; convulsions, irregular heart beat, and hallucinations may be reversed with slow (2 min.) i.v. administration of physostigmine salicylate (2 mg for adults; 0.5 mg for children, repeated every half hour as required, up to 6 mg for adults and 2 mg for children); dexamethasone 1 mg/kg i.v. may be used to reduce fever, or sponge with tepid water (do not give aspirin).

NOTES The name of Jimsonweed for *Datura stramonium* is derived from an incident at Jamestown, Virginia during the Bacon Rebellion of 1676, when soldiers sent to quell the rebellion were poisoned en masse from eating this plant; "Jimson" is a corruption of "Jamestown."

Datura species have been used in many parts of the world since ancient times as hallucinogens and folk medicines. In Europe, for example, Jimsonweed was used to treat mania, epilepsy, melancholy, rheumatism, convulsions, and madness. Recently, Jimsonweed has been an ingredient in powders used to relieve asthma. Thrill-seekers using these powders as hallucinogens and sometimes poisoning themselves through misuse forced the American Food and Drug Administration to ban their over-the-counter sale in 1968.

Mayapple, or American Mandrake
(Barberry Family)

Podophyllum peltatum
(Berberidaceae)

Fig. 80. Mayapple (*Podophyllum peltatum*).
MARY W. FERGUSON

QUICK CHECK Herbaceous plant of open woods and meadows with large, lobed, umbrella-shaped leaves, a single white flower, ripening into a large, yellowish berry. Entire plant, except fully ripe berries, violently purgative; large quantities potentially fatal and can cause birth defects; used in pharmaceutical preparations, but with extreme caution.

DESCRIPTION Perennial herb growing from a fleshy, spreading underground rootstock to a height of about 30–50 cm (12–18 in.). The leaves are large and umbrella-like, with 5–9 prominent lobes spreading out like fingers on a hand, and the stalk joining to the middle of the blade. The leaves are single on flowerless plants, paired on flowering plants. A single large, white, nodding, 5–9 petaled flower is borne between the leaves in early spring. The fruit is a fleshy, yellowish, blotchy berry, ovoid and up to 5 cm (2 in.) long.

OCCURRENCE Often found in large patches in open deciduous woods, wet meadows, and along roadsides in southern Ontario and Quebec and throughout the eastern United States west to Minnesota and Texas.

TOXICITY This plant contains over 15 biologically active compounds, particularly within podophyllin, a resinoid component which includes several lignans, notably podophyllotoxin, and alpha- and beta-peltatin. The ripe berries may be slightly cathartic if eaten in quantity, but are well known as an edible wild fruit. The rest of the plant—rhizome, shoots, leaves, flowers, and unripe fruit—is violently purgative, producing severe digestive upset and diarrhea accompanied by vomiting. It also has a toxic effect on dividing cells, which could potentially lead to birth defects if consumed by women during pregnancy. Children have been poisoned from eating unripe fruit. Poisoning sometimes results from misuse of medicinal preparations of the plant. Eating large quantities of the plant, or repeated application of the resin to the skin may be fatal, producing blood abnormalities, kidney failure, and eventual coma. People handling the powdered rhizomes in commercial drug preparations have experienced eye irritation, keratitis, and ulcerative skin lesions.

Livestock rarely eat this plant, and hence poisoning of them is rare, but deaths do occur occasionally when leaves or shoots are browsed.

TREATMENT Vomiting usually occurs as a symptom, but inducing vomiting or performing gastric lavage for suspected poisoning may be indicated; follow with activated charcoal and a saline cathartic; maintain fluid and electrolyte balance; antiemetic and antidiarrheal agents may be used to control vomiting and diarrhea; blood transfusion may be necessary in severe cases.

NOTES Mayapple has been used in Native folk medicine for centuries, and over 100 years ago found its way into American materia medica. It has been used to treat cancerous tumors, soft warts, and other growths. Unfortunately, the satisfactory use of Mayapple resin against cancer has been complicated by its toxicity.

**This book cannot replace the advice and assistance
of qualified medical personnel.
In all cases of suspected poisoning by plants, or any other substance,
immediate qualified medical advice and assistance should be sought.**

Fig. 81. Milkvetch (*Astragalus miser*). R. & N. TURNER

Milkvetches, or Locoweeds
(Bean, or Legume Family)

Astragalus spp. and
   Oxytropis spp.
(Fabaceae, or Leguminosae)

QUICK CHECK Perennial herbs with pinnately compound leaves, pea-like flowers in elongated clusters, and variously shaped pods. Many types fatally poisonous to animals; leaves, pods, and seeds potentially toxic for humans, especially children who might be attracted to pea-like pods.

DESCRIPTION The Milkvetches, or Locoweeds, are a large, diverse group of plants in two closely related genera. Botanists recognize over 500 species of *Astragalus* in North America alone, and even experts find it difficult to identify them or in some cases to distinguish between members of the two genera. Since both contain toxic species, they are treated here together.

They are perennial herbs, with or without stems, growing in clumps or patches from woody rootstocks. Their leaves are alternate and mostly pinnately compound (with leaflets arranged along a central rib). The flowers, few to many, grow in elongated clusters (racemes) from the leaf axils. They are relatively small and pea-like, varying from white to purple to yellowish depending on the species. In *Astragalus* the flower stems are leafy and the "keel" (lowermost) petal of the flower is blunt, whereas in *Oxytropis* the flower stems are leafless and the keel is prolonged into a distinct point. For this reason, *Oxytropis* is sometimes called Point Locoweed, or Pointvetch. The fruits are pea-like pods varying in size, shape, and texture and containing one or more kidney-shaped seeds.

OCCURRENCE Milkvetches or Locoweeds are found in a wide range of habitats throughout North America, from dry plains, open woods, and hillsides to moist shorelines and alpine meadows.

TOXICITY Some species of Milkvetches are non-toxic, and these are desirable as forage and soil-building plants. However, Milkvetches as a group are recognized as among the most poisonous to livestock in North America. They should be considered potentially poisonous to humans too, especially since they have pea-like pods that may attract children. In most cases, however, large quantities would have to be consumed to be fatal.

Several species of both *Astragalus* and *Oxytropis* are known to cause a disease of grazing livestock known as "loco" (Spanish for crazy) which has been widespread in rangelands of western North America. Typical symptoms of this disease, reflected by its name, include staggering, trembling, and paralysis. The toxic principle involved is not well known. A second type of poisoning is caused by Timber Milkvetch (*Astragalus miser*), a small, delicate plant with fine, small leaflets, scanty pinkish mauve blossoms, and slender pods, and by relatives of this plant. It is common on rangelands of western North America. Symptoms appear relatively quickly after eating the plant and include paralysis with respiratory difficulties—huskiness of voice and coughing, roaring or wheezing. In this case poisoning is apparently due to the presence of miserotoxin, the glycoside of nitropropanol, or glycosides of nitropropionic acid.

TREATMENT If pods and seeds of toxic species are known to have been eaten, induce vomiting or perform gastric lavage; observe closely, and give general supportive treatment for symptoms.

NOTES At least one species of Milkvetch is known to have flower nectar that is poisonous to honeybees, and to have caused serious losses to apiarists in Nevada. Some Milkvetches may accumulate toxic amounts of selenium in regions where this element occurs in the soil.

Several other wild legumes are known to be toxic. One is Buffalo Bean, or False Lupine (*Thermopsis rhombifolia*), a common, yellow-flowered herbaceous perennial of the prairies, which has seeds that have reputedly caused poisoning in children and foliage toxic to livestock. Others, the Lupines (*Lupinus* spp.), are treated in Chapter IV (p. 201).

Milkweeds
(Milkweed Family)

Asclepias spp.
(Asclepiadaceae)

Fig. 82. Milkweed (*Asclepias speciosa*). R. & N. TURNER

QUICK CHECK Large, perennial herbs with leaves in opposite pairs or whorls, attractive white, orange, or purplish flowers in umbrella-like

Fig. 83. Milkweed (*A. speciosa*), ripe pods releasing seeds. R. & N. TURNER

clusters; pointed fruiting pods with silk-tufted seeds; plant exudes milky sap when broken. Well known as livestock poison; potentially toxic to humans when some species are used as wild greens or in herbal preparations.

DESCRIPTION There are over 20 species of Milkweeds in North America, many widespread. It is likely that all species have some degree of toxicity. They are upright, herbaceous perennials up to 1 m (3 ft) or more tall, often branching and growing in large clumps from creeping rhizomes. All parts of the plant exude a milky juice (latex) when broken or injured. The leaves, in opposite pairs or whorls, are broad and elliptical or oblong, or long and narrow, smooth-edged, and up to 20 cm (8 in.) long, often with a prominent central vein. A dense mat of soft, silky hairs may cover both surfaces of the leaves, or the upper side only. The 5-parted flowers are usually fragrant, ranging in color from white to red, orange, or lilac-purple depending on the species. They are attractive and intricate, with 2 series of petal parts, the outer whorl strongly curved back, and the inner modified and horned, and projecting forward to the center of the flower. The flowers are arranged in dense, umbrella-like clusters borne on the upper part of the main stem and branches. The fruits are large, greenish, warty pods, splitting along the side when ripe to expose numerous silky-tufted seeds that are very attractive when they become airborne.

OCCURRENCE Milkweeds are widely distributed in North America, growing from coast to coast, from southern Canada throughout the United States, on dry prairies, pastures, clearings, roadsides, dry streambeds, and lakeshores. Some species grow in wet places, swamps and bogs.

TOXICITY At least some Milkweed species, and probably all to some extent, contain toxic resinoids, notably galitoxin, as well as several cardioactive glycosides and a small amount of an alkaloid. The poisonous effect of the latter is usually masked by the resinoid action.

Although there are apparently no reported fatal poisonings of humans by Milkweed, some species are used as edible wild greens, and should be selected and prepared with caution (see below). There are many cases of livestock poisoning by Milkweed, the symptoms produced by various species differing only in degree. Initial symptoms are depression, weakness, and staggering, followed by prostration, tetanic seizures, labored breathing, high temperature, dilation of the pupils, coma, and death within one to a few days in fatal cases. Fortunately, most Milkweeds are distasteful to livestock, and animals eat them only when no other forage is available.

The milky latex may cause allergic dermatitis in some individuals.

TREATMENT Induce vomiting, or perform gastric lavage; supportive treatment of symptoms.

NOTES Although the young shoots, flowers, and green pods of some species are popular as wild vegetables when cooked, the uncooked shoots and the mature plants should never be consumed. Only known edible species of Milkweeds, such as *Asclepias speciosa* and *A. syriaca,* should be eaten, and then only after cooking the young shoots or pods in one or more changes of water; this eliminates most of the bitter, poisonous compounds, which leach out in the water.

Milkweed species have been used for centuries in folk medicine by Native people and settlers.

Nightshade, Black
(Nightshade Family)

Solanum nigrum and
    S. americanum
(Solanaceae)

Fig. 84. Black Nightshade (*Solanum nigrum*), flowers and green berries. R. & N. TURNER

QUICK CHECK Leafy annuals with white, potato-like flowers and clustered berries, green when unripe and shiny black when ripe. Foliage and green berries highly toxic, sometimes fatal. Fully ripe berries of *some* varieties edible when cooked, but use caution.

DESCRIPTION Branching annuals, erect or sprawling, 15 cm -1 m (6 in.—3 ft) high. The stems and leaves are smooth to roughly hairy. The leaves are alternate, simple, smooth-edged or with blunt teeth, and oval to triangular or lance-shaped. The star-shaped flowers are small, white, and potato-like, with spreading or reflexed petals and yellow stamens grouped tightly around the stigma at the center. The berries, hanging in small clusters, are green when unripe, black and shiny when ripe. There are both native and introduced populations of Black Nightshade. The former are included in *S. americanum*, the latter, *S. nigrum*, but the two species are difficult to distinguish and both are poisonous. Several other species of *Solanum*, also toxic, are found in various parts of North America. All have characteristic potato-like flowers, varying in color from white to blue to violet. [See also Climbing Nightshade (*Solanum dulcamara*) under Wild Vines (p. 94), and, in the same family, Belladonna or Deadly Nightshade (*Atropa belladonna*) in Chapter IV (p. 183).]

OCCURRENCE The Black Nightshades occur as weeds of fields and open woodlands, gardens, pastures, waste places, and disturbed ground throughout North America.

TOXICITY Black Nightshades, like Climbing Nightshade (*Solanum dulcamara*), contain solanine, a toxic alkaloidal glycoside, also present in green potatoes and potato leaves. The mature foliage and green berries are especially toxic. The fully ripe fruits of some varieties are edible when cooked, but toxicity varies, and unless certain the berries are from an edible strain, leave them alone.

Both humans and animals of various types have been poisoned by Black Nightshade, sometimes fatally. Symptoms are similar to those of Climbing Nightshade (see p. 94).

TREATMENT See under Climbing Nightshade (p. 94).

NOTES Another common toxic species of this genus is Horsenettle, or Wild Tomato (*Solanum carolinense*). A tall herbaceous perennial, it is a weed of fields and wastelands, with prickly stems and leaves, and large, yellowish berries which may remain over the winter and cause poisoning even in spring. It is widespread, but especially common in the southeastern United States. The red-berried Jerusalem Cherry (*S. pseudocapsicum*) is another toxic relative (see Chapter V, p. 248).

Plant breeders have developed several edible-fruited varieties from Black Nightshade, variously marketed as "Wonderberry," "Sunberry," or "Garden Huckleberry." These are sometimes classed as separate species: *Solanum burbankii, S. intrusum,* or *S. melanocerasum.*

Peyote Cactus, or Mescal
(Cactus Family)

Lophophora williamsii
(Cactaceae)

Fig. 85. Peyote Cactus (*Lophophora williamsii*).
WALTER H. LEWIS

QUICK CHECK Small, spineless, rounded cactus with one or more stems having rounded sections around the top, a central pinkish flower, and small, pink fruit. Distributed as small, dried "mescal buttons." A potent hallucinogen with potentially dangerous physiological and psychotic effects, but not usually fatal.

DESCRIPTION Peyote is a small, fleshy cactus with one or more short, rounded stems growing from a large, branching perennial rootstock. Each stem, about 2–5 cm (1–2 in.) across, is divided around the top into several low, rounded sections, each bearing a tuft of yellowish white hairs. The flower, white to rose-pink, grows at the center, and ripens into a small, pinkish, berrylike fruit. The seeds are black. The plant is distributed, usually illegally, as an hallucinogen in the form of small, dried "buttons," which are brownish and have a bitter, disagreeable taste.

OCCURRENCE Peyote grows naturally in the desertlands of southern Texas to central Mexico; dried "buttons" are shipped to other parts of North America for use as an hallucinogen. Possession of Peyote, which is classed as a narcotic, is unlawful in some areas, but many places have no legal restrictions against it.

TOXICITY Peyote contains many alkaloids, including mescaline, peyotline, anhaline, anhalamine, anhalanine, and lophophorine. Since the alkaloids act together, and their effects vary with their relative concentration, dosage, and mental and physical state of the person eating peyote, it is difficult to characterize the specific actions of each. However, it is known that lophophorine is the most toxic, having strychninelike effects, whereas mescaline (also known as 3,4,5,-trimethoxy-beta-phenylethylamine) is probably the most hallucinogenic, acting through paralysis of the central nervous system and producing awareness of heaviness of the limbs and color visions. Shortly after ingestion or after a tea from it is drunk, peyote produces nausea, chills, headache, and severe stomach pain and vomiting, often accompanied by terror, anxiety, pupil dilation and visual disturbances, hot, flushed face, muscular relaxation, dizziness, slight decrease in pulse, loss of sense of time, and wakefulness. As these symptoms subside, mental stimulation begins, with clarity and intensity of thought, brilliantly colored, sometimes bizarre, visions, and exaggerated sensitivity to sound and other senses. Peyote apparently does not cause a physiological addiction, but its long-range effects and psychotic reactions are dangerous and it can be psychologically habit-forming. Peyote intoxication usually occurs among young people experimenting with the plant and trying to experience its psychedelic effects.

TREATMENT If thorough vomiting does not occur spontaneously, induce vomiting, or perform gastric lavage; general supportive treatment for symptoms. Unpleasant flashbacks and loss of reality can recur several days after initial recovery.

NOTES Peyote has long been used by Native Indians in religious rites and is still used legally in the communion services of the Native American Church, whose members number about 250,000. Mescaline has also been used as a drug in experimental psychiatry.

Poison Hemlock
(Celery, or Umbel Family)

Conium maculatum
(Apiaceae, or Umbelliferae)

Fig. 86. Poison Hemlock (*Conium maculatum*).
R. & N. TURNER

QUICK CHECK Young plants carrot-like, with basal rosette of finely dissected leaves; mature plants are tall, much branched, with numerous tiny white flowers in umbrella-like clusters; stems smooth, hollow, and purplish-speckled; plants with unpleasant "mousy" odor. Highly toxic and potentially fatal, especially for children, who may mistake the leaves for parsley or the roots for carrots.

DESCRIPTION An erect, branching biennial (sometimes annual or peren-

Fig. 87. Poison Hemlock,
closeup of flowers.
R. & N. TURNER

nial) up to 2 m (6 ft) or more high when mature, growing from a white, fleshy taproot. The leaves, up to 30 cm (1 ft) long, are triangular in outline, finely cut, fern-like, and delicate. The young plant closely resembles Wild Carrot (*Daucus carota*), but the latter has distinctly hairy leaves, whereas those of Poison Hemlock are hairless. The stems are rigid, hollow except at the nodes, smooth, and slightly ridged, with irregular purple blotches or streaks. The flowers are tiny and white, borne in numerous umbrella-like clusters 2–5 cm (1–2 in.) across, distributed from branch tips throughout the upper part of the plant. The fruits are small, grayish brown, and flattened, each with five prominent wavy ridges running lengthwise. When the plant is bruised, crushed, or even touched, a strong, unpleasant odor is emitted, resembling the smell of mice to some people.

OCCURRENCE Originally introduced from Europe, Poison Hemlock has become a common and obnoxious weed of waste ground, pastures and fields, roadsides, ditches, and sometimes gardens, throughout southern Canada and the United States, particularly the northern States and adjacent Canada.

TOXICITY Poison Hemlock contains a group of closely related poisonous alkaloids, including coniine, N-methyl coniine, gamma-coniceine, conhydrine, and pseudoconhydrine. These alkaloids are structurally related to nicotine, and act similarly, producing initial stimulation followed by severe depression of the central nervous system, resulting in paralysis, slowing of the heart, convulsions, and death from respiratory paralysis. All parts of the plant are toxic, especially the leaves before flowering, and the flowers and fruits. The concentration of alkaloids also varies with climatic conditions; sunny summers produce greater quantities.

The plant is known to be toxic to both humans and animals, and has often produced fatalities. In 1980 a 5-year-old girl was killed in Victoria when she ate Poison Hemlock roots, pretending they were carrots. In 1987, another child was seriously poisoned from this plant in Victoria but fortunately recovered. Sometimes people mistake the finely dissected leaves of the young plants for parsley, or the seeds for anise, with fatal results. Symptoms of human poisoning, appearing 1–3 hours or more after eating the plant, include nausea, vomiting, salivation, abdominal pain, diarrhea, headache, dilation of the pupils, lack of coordination, confusion, sweating, salivation, difficulty in breathing, coldness of the extremities, drowsiness, initial high blood pressure followed by low blood pressure, rapid or irregular heart beat, convulsions, and coma. Death usually results from stoppage of breathing. Fortunately, the taste of the plant is so unpleasant that seldom is enough eaten to cause death.

Animals are repelled by the unpleasant odor, but are sometimes poisoned in the spring when the leaves of Poison Hemlock are low and often mixed in with grasses and other forage. Cattle are the most susceptible, but all classes of livestock may be affected. The unpleasant odor of the plant is often detectable in the urine and breath of a poisoned animal. Birth deformities may be caused by pregnant females eating the plant.

TREATMENT Do not induce vomiting or perform gastric lavage if vomiting has occurred or if seizures are imminent; i.v. diazepam may be used to sedate patient, then, as soon as vomiting ceases, administer activated charcoal slurry to adsorb the alkaloids and a saline cathartic (repeat charcoal and cathartic every 6 hours for 24 hours); artificial respiration and oxygen may be required; monitor heart and arterial blood gases; anti-convulsive therapy, with i.v. diazepam or phenobarbital is suggested; general supportive treatment for symptoms.

NOTES To prevent the possibility of poisoning of children and animals, eliminate Poison Hemlock plants from yards, playgrounds, and pasturage before the flowers have a chance to go to seed.

It was this plant that was used in 399 B.C. to kill the Greek philosopher Socrates; forcing the victim to drink a Poison Hemlock solution was apparently a widely used means of execution by the early Greeks.

The most notorious toxic relative of Poison Hemlock is the violently poisonous Water Hemlock (*Cicuta* spp.) (see p. 141). Several other species in the family have been implicated in poisoning, including some species of Water Parsnip (*Sium* spp.) (generally considered edible; cf. Turner 1978), Cow Parsley (*Anthriscus sylvestris*), and Fool's Parsley (*Aethusa cynapium*). Another relative, Water Parsley (*Oenanthe sarmentosa*), is implicated by the known toxicity of a European species of this genus, Hemlock Water Dropwort (*Oenanthe crocata*).

Poison Hemlock and Water Hemlock should not be confused with the native coniferous tree called Hemlock (*Tsuga* spp.), which is not poisonous.

Pokeweed, or Pokeberry
(Pokeweed Family)

Phytolacca americana
(Phytolaccaceae)

Fig. 88. Pokeweed (*Phytolacca americana*). FRANK FISH

QUICK CHECK Tall, large-leaved herb with whitish flowers and round, purple-black berries borne in elongated clusters. Entire plant, especially raw berries, highly toxic and potentially fatal. Even young shoots, considered edible by some, *should not be* used due to the presence of a blood cell altering chemical; poisoning sometimes occurs from misuse of herbal preparations.

DESCRIPTION Large perennial herb up to 2.5 m (8 ft) or more tall, growing from a thick, fleshy taproot, with greenish to reddish branching stems. The leaves are alternate, oblong, and smooth-edged, those near the base as much as 30 cm (1 ft) long. The many small flowers, with greenish white to pink sepals, are borne in an erect or drooping cluster (raceme) up to 20 cm (8 in) long. The fruits are round, shiny, dark purple, juicy berries.

OCCURRENCE Common native plant of rich, disturbed soils; found in open fields, moist woods, and waste places and along fencerows and roadsides throughout eastern United States and southeastern Canada; occasional weed on the West Coast. Sometimes grown in gardens. A related species, *Phytolacca rigida*, with upright fruit clusters, occurs on coastal dunes and marshes from North Carolina to Texas and inland in Florida.

TOXICITY The entire plant, especially the roots and seeds, is highly poisonous. Several toxic compounds are present, including water-soluble triterpene saponins, one of which has been identified as phytolaccigenin. Two others are phytolaccin and phytolaccatoxin. Poisoning symptoms, generally developing 1–2 hours after eating Pokeweed, include abdominal cramps, persistent vomiting, diarrhea, and, in severe cases, convulsions and death. Salivation, heavy

Fig. 89. Pokeweed fruits. WALTER H. LEWIS

perspiration, weakness and drowsiness, slowed pulse, difficult breathing, and visual disturbance may also be experienced. An immediate burning sensation in the mouth is the first warning, and fortunately usually stops people from eating a fatal dose. Most fatalities occur in children who eat the berries. Infants and toddlers can be seriously poisoned from eating only a couple of raw berries. One 5-year-old died from drinking a quantity of crushed pokeberries added to sugar and water to simulate grape juice. Animals find Pokeweed distasteful and do not usually eat it, although pigs are sometimes poisoned from eating the roots, and the plant is a potential hazard to all classes of livestock.

Probably the most dangerous property of Pokeweed is that, as has been recently found, it contains a type of protein lectin, a mitogen, that can cause serious and wide-ranging blood cell abnormalities. The Pokeweed mitogen affects division of human white blood cells, and induces the proliferation of B and T lymphocytes. The mitogen can be absorbed through cuts and skin abrasions, as well as through ingesting the plant. Therefore, Pokeweed *should not be* handled except with gloves. Gardeners and others trying to eliminate the plant should be particularly careful.

TREATMENT Induce vomiting, or perform gastric lavage; supportive treatment for symptoms. The Pokeweed mitogen may disturb the body's immune system, and this effect should be watched for in cases of Pokeweed poisoning.

NOTES The young leafy shoots of Pokeweed are frequently used as a springtime cooked vegetable in eastern North America. The cooked berries have also been eaten. Although thorough cooking (in two waters for the shoots) eliminates most of the toxins, the danger of absorbing the cell-altering mitogens during harvesting and preparation of the shoots and berries is very real. Furthermore, people may be poisoned from improperly cooked leaves or from roots pulled up with the shoots. Because of the potential dangers involved, people should avoid using Pokeweed altogether. Folk medicinal preparations from Pokeweed roots are also potentially dangerous.

Ragwort, or Groundsel
(Aster, or Composite Family)

Senecio jacobea and related
    species
(Asteraceae, or Compositae)

Fig. 90. Ragwort (*Senecio jacobea*). R. & N. TURNER

QUICK CHECK Tall, weedy plant of pastures and roadsides with irregularly lobed leaves, and yellow flowers in conspicuous flat-topped clusters. Toxic and potentially carcinogenic; may produce severe liver damage; seldom immediately fatal in humans, but may be a harmful contaminant of milk and honey. Do not use in herbal preparations or teas.

DESCRIPTION *Senecio jacobea,* also known as Tansy Ragwort, is a tall, coarse herbaceous biennial, winter annual, or occasional perennial, up to 1 m (3 ft) or more high, with tough upright stems often red-tinged near the base and branching above the middle. A basal rosette of leaves usually dies before flowering, but the stem leaves persist. They are deeply and irregularly lobed or divided, dark green, and tough. The composite (daisy-like) flowers are small, bright yellow, and numerous, borne in dense, conspicuous, flat-topped clusters. The single-seeded fruits are topped with a downy "parachute," making them readily dispersable. There are many other species of Ragwort, or Groundsel, and over 25 of them have proven poisonous. Some, such as Common Groundsel (*S. vulgaris*), are weedy annuals; others (e.g., *S. integerrimus*) are native perennials. In cases where Ragwort poisoning is suspected, a botanist should be consulted for accurate identification, because the genus *Senecio* is complex and difficult to characterize.

OCCURRENCE Introduced from Europe, Tansy Ragwort is now a common weed of roadsides, fields, and pastures throughout much of North America, particularly in the Atlantic and Pacific coastal regions. It has taken over entire fields of previously productive pastureland.

TOXICITY Most Ragworts should be considered potentially poisonous, but Tansy Ragwort (*Senecio jacobea*) is the most notorious, and is considered in Britain to be one of the most important of all poisonous plants. It contains a wide spectrum of pyrrolizidine alkaloids, including senecionine, seneciphylline, jaconine, and jacobine. The most toxic are cyclic diesters. They are apparently metabolized in the liver to bound pyrrole derivatives, both soluble and insoluble, that are highly toxic and carcinogenic. The toxins are not destroyed by drying or storage. In some parts of the world, such as Africa and the West Indies, humans have suffered from chronic Ragwort poisoning caused by eating bread made from flour contaminated with seeds of Ragwort species or by drinking medicinal "tea" from some species. Symptoms of human poisoning include abdominal pain, nausea, vomiting, headache, enlarged liver, apathy, and emaciation. Additionally, people may be harmed indirectly from Ragwort-tainted milk or honey. Some Ragwort species also cause contact dermatitis.

Grazing livestock are frequently affected by Ragwort. Field poisoning of livestock usually occurs only when there is a heavy infestation of the plants and when other forage is scarce. Cattle and horses are the most frequent victims. Symptoms of Ragwort poisoning—digestive disturbances, restlessness, lack of coordination, and paralysis—may develop from a few days to several weeks or months after animals have eaten the plants. Once symptoms are obvious, an animal may die within a few days.

TREATMENT There is no specific treatment for Ragwort poisoning, only general supportive treatment for symptoms of digestive distress and liver damage, and, of course, removing the source of poisoning.

NOTES Ragwort poisoning of animals was recognized only within the present century and was first determined, after great difficulty, in 1906 in Nova Scotia. There a condition known as Pictou disease, after the town of that name, had become common following the introduction of Ragwort from Scotland. Today in Britain Ragwort poisoning is said to cause higher economic losses in cattle than all other plants combined, and the spread of Ragwort is carefully monitored and controlled through mechanical removal and spraying with herbicides.

Several plants in the borage family (Boraginaceae), including Heliotropes (*Heliotropium* spp.), Comfrey (*Symphytum* spp.), Hound's Tongue (*Cynoglossum officinale*), Viper's Bugloss (*Echium* spp.), and Fiddleneck, or Tarweed (*Amsinckia intermedia* and related species), as well as Rattlebox (see following), also contain liver-damaging pyrrolizidine alkaloids, and *should not be* used internally as medicinal herbs, even though some have been in the past.

Fig. 91. Rattlebox (*Crotalaria spectabilis*).
MARY W. FERGUSON

Rattlebox, or Crotalaria
(Bean, or Legume Family)

Crotalaria spectabilis and related
   species
(Fabaceae, or Leguminosae)

QUICK CHECK *Crotalaria spectabilis* is a dense herbaceous annual up to 2 m (6 ft) tall, with large, simple leaves, yellow, pea-like flowers in showy clusters, and inflated pods whose seeds rattle inside when dried. Entire plant highly toxic, producing severe liver damage; seeds and "medicinal" tea from leaves are the usual causes of human poisoning.

DESCRIPTION There are several species of Rattlebox growing in parts of North America, most of them coarse, yellow-flowered herbs with distinctive, inflated pods that rattle when shaken. In terms of toxicity, *Crotalaria spectabilis* is the most important and widely distributed. It is a dense, erect annual (sometimes perennial, but never woody), growing up to 2 m (6 ft) high, with smooth, erect stems. The leaves are simple and oval, up to 18 cm (7 in.) long, hairless above and finely hairy beneath. The flowers are large (2.5 cm or 1 in. long), and yellow, with purple tinges or veins, borne in large, elongated terminal clusters. The fruiting pods, up to 5 cm (2 in.) long, are inflated, black at maturity, and contain about 20 glossy black seeds which detach at maturity and rattle when the pod is shaken.

OCCURRENCE *C. spectabilis* is a common weed of fields and roadsides from the southern states north to Missouri and Virginia; occasionally cultivated as an ornamental or soil conditioner. Various other species of *Crotalaria* occur as garden plants or introduced weeds in parts of the United States. One native species, *C. sagittalis,* is found in scattered populations in valley bottoms of the southern and central states.

TOXICITY Like the Ragworts (*Senecio* spp.) and some members of the borage family, Rattlebox species contain a dangerous group of pyrrolizidine alkaloids in their seeds, leaves, and stems. One of the major alkaloid compounds isolated from Rattlebox is called monocrotaline. Humans have been poisoned through eating the seeds (as a contaminant of grain) or drinking a "medicinal" tea made from the plant. The main effect is severe liver damage (veno-occlusive liver disease, known as Budd-Chiari syndrome, with hepatic vein thrombosis leading to cirrhosis). Symptoms are abdominal pain, enlargement of liver and spleen, loss of appetite, nausea, vomiting, and diarrhea.

Livestock of all types, particularly horses, cattle, fowl, and swine, have often been poisoned by *C. spectabilis* and other Rattlebox species. Animal poisoning generally parallels that of humans, with severe damage to liver and spleen.

TREATMENT There is no specific treatment for Rattlebox poisoning; general supportive therapy for symptoms of digestive upset and liver damage.

NOTES *Crotalaria spectabilis* and several other Rattlebox species were originally introduced to the United States in 1921 by the Bureau of Plant Industry as potential soil builders, hay, and forage crops; it was not until a decade later that their toxicity was discovered.

Another toxic plant in the same family, *Daubentonia punicea*, is also sometimes called Rattlebox, as well as False Poinciana or Sesbane. Its seeds and flowers contain toxic saponins and it sometimes occurs as a garden escape in the southern United States [see also under Poinciana (*Poinciana gilliesii*) in Chapter V, p. 235].

St. John's-Wort, or Klamath Weed
(St. John's-Wort Family)

Hypericum perforatum
(Hypericaceae)

Fig. 92. St. John's-Wort (*Hypericum perforatum*).
R. & N. TURNER

QUICK CHECK Erect, opposite-leaved perennial herb; leaves spotted with tiny translucent dots; flowers bright yellow and numerous, in terminal clusters. Entire plant contains a phototoxin which can cause dermatitis and inflammation of the mucous membranes on exposure to direct sunlight; use extreme caution if taking as medicinal herb.

DESCRIPTION An perennial herb up to 1 m (3 ft) or more tall, with winged or 2-edged stems and branches. The stems are smooth and erect, with a woody base. The leaves are opposite, lacking stalks, elongated and elliptical, with pointed tips, and up to 2.5 cm (1 in.) long. They are spotted with many tiny glandular dots; when held up to the light, these appear translucent. The flowers are 5-petaled, yellow, with numerous prominent yellow stamens, and are borne in flat or round-topped clusters towards the top of the plant. The fruiting capsules, brown when mature, contain numerous small pitted seeds.

OCCURRENCE St. John's-Wort was introduced to the eastern United States many years ago, and is now an aggressive weed of roadsides, pastures, and ranges throughout most of the United States and southern Canada.

TOXICITY St. John's-Wort and most other *Hypericum* species contain hypericin, a reddish, fluorescent substance believed to be a naphthodianthrone derivative. Concentrated in the glandular dots on the leaves, it is a phototoxin, a poisonous or irritant substance which is activated by sunlight, specifically, ultraviolet rays. It causes swelling, blistering, and lesions on the unprotected skin of animals consuming the plant, particularly of albinos and others of light coloring, and can cause similar reactions in humans ingesting the plant in herbal medicine preparations. Once sensitivity to hypericin is developed, subsequent reactions to eating the plant or its extracts and exposure to sunlight become more and more serious.

Seriously affected animals may also exhibit thrashing of limbs, loss of appetite, diarrhea, increased respiration and heartbeat, high temperature, blindness, staggering, convulsions, and sometimes coma and death. Depending on conditions, symptoms may appear from 2 to 14 days after an animal has eaten St. John's-Wort. Sometimes an animal may survive the acute stages of poisoning, but die eventually from secondary reaction, such as refusal to eat.

TREATMENT Treat for symptoms of dermatitis, and do not allow further contact or ingestion of plant. Avoid infection of open skin lesions.

NOTES St. John's-Wort is common and widespread, but it was formerly much more noxious. Introduced into California around 1900, Klamath Weed, as it was soon named, had spread by the 1950s over an estimated 2⅓ million acres in that state alone, and was considered the worst cause of economic loss of pasture and rangelands of California. It caused stock fatalities both directly and indirectly, by reducing the available areas of useful grasses and forage. Fortunately, a biological control program using beetles that feed specifically on it has brought a dramatic decrease in its spread.

There are many other species of *Hypericum* in North America, both native and introduced. Some are grown as garden ornamentals. The same phototoxic compound is known or suspected in some of these, but *H. perforatum* is by far the most notorious.

Snakeroot, White
(Aster, or Composite Family)

Eupatorium rugosum
(Asteraceae, or Compositae)

Fig. 93. White Snakeroot (*Eupatorium rugosum*).
WALTER H. LEWIS

QUICK CHECK Tall, perennial herb with long-stalked, toothed leaves, and small, white flower heads in terminal clusters. Entire plant highly toxic; can contaminate milk of cows browsing it and cause severe, potentially fatal, illness in humans; with modern dairy milk production "milk sickness" is almost unknown today, but people owning milk cows should beware.

DESCRIPTION This plant, also called Fall Poison, is a showy, erect perennial herb, with branched or unbranched stems, up to 1.2 m (4 ft) tall. The leaves, 8–15 cm (3–6 in.) long, are paired and opposite, long-stalked, oval to heart shaped, thin, strongly 3-ribbed beneath, sharply toothed at the edges, and pointed. The composite flower heads are white, and borne in open, rounded clusters at the top of the plant. Fruits are small achenes with parachute-like tufts of hair to aid in dispersal. There are many other species of white-flowered *Eupatorium,* and this species is itself highly variable. Therefore, if poisoning is suspected, a botanist should be asked to confirm the identification. Only *E. rugosum* is implicated as being toxic.

OCCURRENCE A common plant of rich, open woods and recently cleared areas from eastern Canada to Saskatchewan, and south to eastern Texas, Louisiana, and Georgia. It typically grows in low, moist wooded areas, or along streams, draws, and ravines.

TOXICITY The whole plant contains a highly toxic complex alcohol, tremetol, in combination with a resin acid and several glycosides. It was responsible, especially in the early 1800's, for many human deaths from North Carolina to the midwestern states, from a condition called "milk sickness." Cows eating large quantities of this plant would become sick with a disease called "trembles." The milk they produced contained high concentrations of the poisonous substance. People drinking the milk developed symptoms including weakness, nausea, vomiting, tremors, jaundice, constipation, prostration, delirium, and in severe cases, death. Fortunately, with a better understanding of the cause of poisoning, and because most milk is now produced commercially, "milk sickness" is rare today. White Snakeroot, fresh or dried, is also toxic to other animals, and to humans if they consume it directly. Symptoms of poisoning include sluggishness, difficulty in walking, salivation, loss of appetite, and labored breathing, as well as the characteristic trembling and other symptoms of "milk sickness."

TREATMENT Supportive therapy for symptoms; treat for liver damage and urine retention. Remove milk cows from any pasture where White Snakeroot occurs, and do not use milk from cows that have been eating this plant.

NOTES "Milk sickness" from White Snakeroot was said to have been the cause of death of Abraham Lincoln's mother. A related plant, Rayless Goldenrod, or Jimmy Weed (*Haplopappus heterophyllus*), is also known to contain tremetol, and has also caused "milk sickness." It is an erect, bushy plant with slender, sticky leaves, and small, yellow flower heads in terminal clusters. It is common in fields, rangelands, and wet areas, from the central United States to Arizona and New Mexico.

There are many other plants in the aster, or composite family that are toxic to livestock and potentially harmful to humans if used as herbal preparations or if they contaminate milk; it is not possible to describe all of them in a book such as this. Interested readers should consult Kingsbury (1964) and Fuller and McClintock (1986) for detailed listings. Some notable poisonous composites are: Bitterweed (*Hymenoxys* spp.); Broomweed (*Gutierrezia microcephala*); Cloth of Gold (*Baileya multiradiata*); Cone Flower (*Rudbeckia laciniata*); Copperweed (*Oxytenia acerosa*); Goldenrods (*Solidago* spp.); Horsebrush (*Tetradymia* spp.); Paperflower (*Psilostrophe* spp.); Sartwellia (*Sartwellia flaveriae*); Sneezeweed (*Helenium hoopesii* and related species); Tarbush, or Blackbrush (*Florensia cernua*); Yellow Star Thistle (*Centaurea solstitialis*); and Stinking Mayweed, or Dog Fennel (*Anthemis cotula*). Other toxic composites discussed elsewhere in this book include: Cocklebur (*Xanthium strumarium*) (p. 105); Ragwort, or Groundsel (*Senecio* spp.) (p. 133); and Tansy (*Tanacetum vulgare*) (p. 140).

Tansy, Common
(Aster, or Composite Family)

Tanacetum vulgare (syn.
Chrysanthemum vulgare)
(Asteraceae, or Compositae)

Fig. 94. Common Tansy (*Tanacetum vulgare*).
R. & N. TURNER

QUICK CHECK Tall, strong-scented perennial herb with deeply toothed leaves, and yellow, button-like flower heads in flat-topped clusters. Entire plant toxic, causing serious illness and occasional fatalities when misused as herbal tea, flavoring, or medicinal "oil"; use in herbal medicine not recommended.

DESCRIPTION Erect herbaceous perennial up to 1 m (3 ft) tall, often growing in thick patches. The leaves and stems are strongly aromatic when crushed or bruised. The leaves are dark green, alternate, and pinnately divided into narrow, deeply toothed segments. The flower heads are bright yellow and button-like, without rays, and borne in large, flat-topped, terminal clusters.

OCCURRENCE Originally introduced from Europe as a garden herb, Common Tansy is now a widespread weed of roadsides, pastures, and waste places throughout North America. Still grown in gardens, it is also marketed as an herbal tea.

TOXICITY Tansy contains a volatile oil, tanacetin or "oil of tansy," which is extremely variable in its chemical composition, but usually contains large amounts of thujone. The dried leaves and flowering tops, in the form of a herbal tea, have been used to kill intestinal worms, to promote menstruation in women, and even to cause an abortion. Use of this tea, however, is not recommended, both because its chemical content is so variable (some chemical races contain no thujone; others up to 95% thujone in their oil), and because thujone itself is a relatively toxic compound that can cause convulsions and psychotic effects in people. Overdoses of this tea, or of "oil of tansy," have caused serious poisoning and have, on occasion, been fatal. Symptoms of Tansy poisoning include rapid, feeble pulse, severe inflammation of the stomach lining, violent spasms, and convulsions. Tansy is also potentially poisonous to grazing livestock and is suspected to have caused abortion in cattle, but because of its strong taste and smell it is rarely touched by animals.

TREATMENT Induce vomiting, or perform gastric lavage; treat for symptoms.

NOTES Several related plants also containing thujone in their aromatic oils have been used in herbal medicine. These include Yarrow (*Achillea millefolium*), and various Sagebrush and Wormwood species (*Artemisia* spp.), especially Absinthe (*Artemisia absinthium*), the main flavoring ingredient in a now-banned liqueur of the same name. This alcoholic beverage, popular at the turn of the century in France, was blamed for profoundly affecting the mental and physical state of those under its influence. Scientists now believe that the thujone in the absinthe is similar to Marijuana in its effects on the mind; hence, plants containing thujone should be regarded with extreme caution.

Water Hemlock, or Cowbane
(Celery, or Umbel Family)

Cicuta spp.
(Apiaceae, or Umbelliferae)

Fig. 95. Water Hemlock (*Cicuta douglasii*).
R. & N. TURNER

QUICK CHECK Erect herbaceous perennials of damp ground and shallow water; thick, fleshy underground stem base is divided internally by cross partitions, with a series of hollow chambers; leaves compound, with many toothed segments, and leaf veins in most species directed to the "V" between the teeth, rather than the tips; white flowers in umbrella-shaped clusters. Considered the most poisonous plant genus in North America. Entire plant, especially roots and rootstock, highly toxic, often fatal if ingested; poisoning occurs rapidly. Sometimes mistaken for edible look-alikes, with deadly results.

DESCRIPTION There are several species of Water Hemlock in North America, all sharing similar diagnostic features, and all very toxic. They are perennial herbs growing from a thickened, fleshy stem base or rootstock, which, when cut open lengthwise, is shown to be

Fig. 96. Water Hemlock,
closeup of leaves.
R. & N. TURNER

divided into chambers by a series of cross partitions. These chambers are usually hollow, but in young plants they may be less obvious. A number of fleshy, tuber-like roots are clustered around the rootstock. The entire plant has a parsnip-like odor which is pungent but not unpleasant. A yellow juice (the toxin itself) exudes from the white flesh of the rootstock when cut. The stems, 1–2 m (3–7 ft) tall when mature, are often purple-striped or mottled, and are hollow except for cross partitions at the leaf nodes. The leaves are alternate, with stalks clasping the stem, and 2–3 times divided into many narrow, toothed leaflets, each usually 2.5–10 cm (1–4 in.) long. In most species (except *C. bulbifera* and *C. californica*) the major leaf veins are directed toward the notches between the teeth, not at the tips as in other plants of the same family. The flowers are small and white, in umbrella-shaped clusters, as is typical for members of this family. There are often many flower clusters on a single plant, borne on branches at or near the top. The fruits are small, ribbed, and paired.

OCCURRENCE Various species are found throughout North America, in shallow ditches, wet meadows, lake and pond edges, marshes, and slow-running creeks. Common species include: Spotted Water Hemlock (*Cicuta maculata*) found from Alaska across Canada and the eastern United States; *C. bulbifera* (which bears clusters of bulblets in the axils of its upper leaves), found across Canada and the northern States; and Douglas' Water Hemlock (*C. douglasii*) of western North America, from Alaska to southern California and Mexico.

TOXICITY Some authorities consider Water Hemlock to be the most violently poisonous plant of the North Temperate Zone; ingestion of one portion of root the width of a finger can be fatal to an adult. All parts of the plant are toxic, especially the fleshy rootstock and roots. The toxin, a yellowish, oily liquid which exudes from the rootstock and roots when cut, is a complex, highly unsaturated alcohol known as cicutoxin, which acts on the central nervous system. Human poisoning has been quite common, and usually occurs through con-

Fig. 97. Water Hemlock, cross-section of chambered rootstock, showing yellow exudation of the poisonous compound, cicutoxin.
R. & N. TURNER

fusing Water Hemlock with similar looking edible species. Many edible plants, including Parsnip, Carrot, Celery, and wild edible roots such Water Parsnip (*Sium* spp.), and Wild Caraway, or Yampah (*Perideridia* spp.) are in the same family. Children have been poisoned from using the hollow stems for peashooters or whistles.

Symptoms of poisoning occur from 15 minutes to one hour from the time of ingestion, and include nausea, salivation and frothing at the mouth, vomiting, violent convulsions and spasms, fever, low heart rate, possible low blood pressure, tremors, severe abdominal pain, widely dilated pupils, delirium, coma, respiratory paralysis, and death, which often occurs before medical help can even be sought. If the patient survives the first few hours, he will usually recover.

Water Hemlock is equally poisonous to all types of livestock, and has caused severe losses of grazing animals which may eat exposed rootstocks from cleared out ditches or the dried edges of lakes and ponds. The leaves and stems are also sometimes browsed but are not as toxic.

TREATMENT   Do not attempt to induce vomiting, or perform gastric lavage; sedate patient with i.v. diazepam, then administer activated charcoal and a saline cathartic (repeat charcoal and cathartic every 6 hours for 24 hours in severe cases); keep patient warm and avoid excitement; control convulsions with i.v. diazepam or phenobarbital, supplemented with pancuronium if necessary; keep air passage open; artificial respiration may be required; monitor blood (arterial gases, pH), temperature, heart rate, kidney function; maintain fluid and electrolyte balance; correct acidosis with i.v. sodium bicarbonate; assure adequate urine flow with mannitol and furosemide as necessary; hemodialysis may be indicated; observe patient closely for 24 to 36 hours, until symptoms completely disappear.

NOTES   Native peoples of North America have long been aware of the poisonous qualities of Water Hemlock. A well known Indian antidote—said to be the only effective one for this and other highly poisonous plants—is feeding the victim salmon oil skimmed from salmonhead soup.

A toxin closely related to cicutoxin is found in another wetland plant, Water Parsley (*Oenanthe* spp.), also of the celery family. A European species of this genus, *O. crocata,* known as Hemlock Water Dropwort, has caused fatalities of humans and livestock in Britain and elsewhere. It has been introduced around Washington, D.C. Other *Oenanthe* species should be considered potentially toxic.

Another toxic relative, often confused with Water Hemlock, is the introduced weed, Poison Hemlock (*Conium maculatum*), which is also described here (p. 128).

References for wild trees, shrubs, vines and flowers; Chapter III: Baer 1979; Baginski and Mowszowicz 1963; Bruce 1927; Canada Department of Agriculture 1968; Canadian Pharmaceutical Association 1984; Claus et al. 1970; Cooper and Johnson 1984; DeWolf 1974; Fernald et al. 1958; Fuller and McClintock 1986; Hardin and Arena 1974; Hill 1986; Howard 1974; Keeler 1979; Keeler et al. 1978; Kingsbury 1964; Krochmal et al. 1974; Lampe and McCann 1985; Levy and Primack 1984; Lewis 1979a, 1979b; Lewis and Elvin-Lewis 1977; McLean and Nicholson 1958; McPherson 1979; Meijer 1974; Mitchell and Rook 1979; Morton 1958, 1971; National Safety Council 1975; Schneider 1986; Soper and Heimburger 1982; Sweet 1962; Turner 1975, 1978, 1984; Tyler 1987.

CHAPTER IV

POISONOUS GARDEN AND CROP PLANTS
(Including Common Garden Weeds)

This book cannot replace the advice and assistance
of qualified medical personnel.
In all cases of suspected poisoning by plants, or any other substance,
immediate qualified medical advice and assistance should be sought.

TREES AND TALL SHRUBS

Allamanda, Yellow
(Dogbane Family)

Allamanda cathartica
(Apocynaceae)

Fig. 98. Yellow Allamanda (*Allamanda cathartica*).
MARY W. FERGUSON

QUICK CHECK Showy, yellow-flowered shrub or vine of warm regions;
  may cause vomiting and diarrhea, but not known to be fatal.
DESCRIPTION Woody climber, often pruned to a somewhat sprawling
  shrub up to 5 m (15 ft) tall. The leathery, glossy leaves are large (up to

145

15 cm or 6 in. long) and lance-shaped or elliptical, opposite or sometimes in whorls of 3 or 4. The flowers, clustered near the ends of the
branches, are yellow, showy, tubular, and trumpet shaped, with 5
flaring petal lobes. The fruit is a prickly capsule.

OCCURRENCE Yellow Allamanda, a native of Brazil, is a popular garden
ornamental of the tropics and subtropics and is commonly cultivated
in the southern United States and Hawaii.

TOXICITY All parts of the plant, especially the fruit and seeds, and the sap
of the stems and leaves, cause mild to severe stomach upset and
vomiting; hence, the specific name, *cathartica.* The cathartic principle
itself is unidentified.

TREATMENT Induce vomiting if it does not occur spontaneously, or perform gastric lavage; treat for symptoms; check for dehydration, especially in young children.

NOTES Yellow Allamanda is often propagated by cuttings, which do not
fruit. This eliminates a common source of poisoning—the fruits and
seeds. A related species, *A. neriifolia,* is cultivated over the same
general area. It has smaller leaves and flowers, and fruits frequently.

Apricot
(Rose Family)

Prunus armeniaca
(Rosaceae)

Fig. 99. Seed kernels, leaves, and bark of Apricot,
Plum, Peach, and Apple all contain cyanide-
producing compounds and can be deadly if eaten in
large quantities. R. & N. TURNER

QUICK CHECK Deciduous tree with pinkish, early blossoms, simple, oval-
shaped leaves, and fleshy, yellowish to orange fruits each containing a
large, ridged stone; fruits edible, but bark, leaves, and seed kernels are
cyanide-producing; potentially fatal if swallowed in quantity.

DESCRIPTION Small, deciduous tree with reddish bark and smooth twigs.
The leaves are oval, about 5 cm (2 in.) long, finely toothed at the
edges, and with a pointed tip. The flowers, appearing before the
leaves, are pinkish and up to 2.5 cm (1 in.) across. The fruits, widely
marketed, are fleshy and yellowish to orange, often flashed with red.
Smaller than peaches, they are slightly flattened, with a characteristic

groove along one side. Each fruit contains a large, single, brown stone, flattened and pointed at each end and grooved along the sides. The hard stone encloses one or two almond-like seed kernels.

OCCURRENCE Apricot, its larger relative, Peach, and hybrid Nectarine are all common and important cultivated fruits in North America. There are many forms, some grown for their fruits, others for their blossoms. Apricot pits are sometimes sold in health food stores as herbal medicines, but often without proper labelling as to their dangers.

TOXICITY Several of our common fruiting trees, including Apricot, Peach and Nectarine (*Prunus persica*), Cherries and Plums (*Prunus* spp.), Apples (*Malus* spp.), and Pears (*Pyrus* spp.) contain substantial amounts of the cyanide-producing glycoside, amygdalin, in their bark, leaves, and seeds or pits. Poisoning symptoms are difficult breathing, inability to speak, twitching, spasms, and in severe cases, coma and sudden death. Children have been poisoned from swallowing the seeds or seed kernels of these species, chewing on the twigs, or making "tea" from the leaves.

TREATMENT If the leaves, bark, or pits of Apricot or the other trees mentioned have been eaten in any quantity, or if nausea occurs, treat for cyanide poisoning (see p. 12): inhalation of amyl nitrate to dilate blood vessels; i.v. injection of sodium nitrite, followed with i.v. sodium thiosulfate, to combine with hydrocyanic acid forming nontoxic thiocyanate which is readily excreted. There may be no time to induce vomiting before the cyanide treatment must be implemented.

NOTES A number of serious cases of poisoning have been reported from misuse of Apricot seed kernels or pits as a source of the drug called laetrile. This compound, widely promoted and used in the 1970s as a cure for cancer, is ineffective and dangerous, without any scientific proof of efficacy, according to Tyler (1987).

Besides the cyanide-containing seeds of various *Prunus* and *Malus* species, several other plants in the rose family have seeds and foliage which should be considered suspect: Saskatoon or Service Berry (*Amelanchier alnifolia* and related species); Mountain-Ash, or Rowan Tree (*Sorbus* spp.), Hawthorns (*Crataegus* spp.), Pyracantha (*Pyracantha* spp.), and Cotoneaster (*Cotoneaster* spp.). These seldom cause poisoning, but should be treated with caution. Saskatoon berries are a widely used edible fruit in many parts of North America, and Mountain-ash and Hawthorn berries, though not very palatable, are sometimes used to make jelly. Berries of another shrub in the same family, Jetberry (*Rhodotypos tetrapetala*), on one occasion caused severe poisoning in a child, simulating a diabetic coma, with severe low blood sugar, fever, and convulsions. The child was given fever-reducing medication and anticonvulsants, and recovered within five days.

**Black Locust, Honey Locust, or
False Acacia**
(Bean, or Legume Family)

Robinia pseudo-acacia
(Fabaceae, or Leguminosae)

Fig. 100. Black Locust (*Robinia pseudo-acacia*).
R. & N. TURNER

QUICK CHECK Deciduous tree with furrowed bark, compound leaves, whitish (or pink), fragrant flowers in hanging clusters, and bean-like seed pods. Entire plant potentially highly toxic from lectins, but human fatalities unknown.

DESCRIPTION Deciduous tree up to 25 m (80 ft) tall, with deeply furrowed, dark brown bark and thorny branches. The leaves are alternate and pinnately compound, with 7–19 leaflets, arranged on either side of a central vein, or axis. The leaflets are oval or elliptical and up to 5 cm (2 in.) long. The flowers are whitish (pink in some varieties), up to 2 cm (0.75 in.) long, and very fragrant, growing in dense, hanging clusters. The fruits are flattened, hanging, bean-like pods, each containing several seeds. When mature they are dark brown to blackish and often remain on the branches over the winter.

OCCURRENCE This tree is native to the woods of the eastern United States from Pennsylvania to Georgia and west to Iowa, Missouri, and Oklahoma. It is widely planted in gardens and along boulevards throughout temperate North America as a shade and ornamental tree. There are many varieties in cultivation, some with golden or purplish foliage, others with pinkish flowers.

TOXICITY The entire tree, particularly the bark, leaves, and seeds, contains poisonous proteins similar to ricin in Castor Bean, but considerably less toxic. Known as lectins, or toxalbumins (robinin and phasin), they interfere with protein synthesis in the small intestine. A glycoprotein, either robinin or some other substance, which agglutinates red blood cells, has also been extracted from the plant. The flowers are the least toxic. Human poisoning from Honey Locust is potentially serious, but fortunately rare, and fatalities are apparently unknown. Symptoms appear after several hours, and include lassitude, nausea, vomiting, abdominal pain, diarrhea, loss of appetite, dilation of pupils, delirium, confusion, stupor, seizures, diabetes, and in severe cases, respiratory depression, coma, and circulatory collapse, which can occur from two to several days after ingestion.

Fig. 101. Black Locust, hanging seedpods.
R. & N. TURNER

There have been many instances of serious, even fatal poisoning in animals.

TREATMENT Induce vomiting, or perform gastric lavage; follow with activated charcoal and a saline cathartic, repeated every 6 hours for 24 hours; supportive treatment for symptoms; antacids may provide relief; correct fluid and electrolyte balance and maintain ventilation. Convulsions may be treated with i.v. diazepam, and low blood sugar with insulin.

NOTES The most notorious case of human poisoning occurred in 1887, when 32 boys at a Brooklyn orphanage, for reasons unknown, ate the inner bark of Black Locust fence posts from the yard of the institution. Two boys were severely poisoned but all eventually recovered.

Do not confuse Black Locust, sometimes called Honey Locust, with the true Honey Locust (*Gleditsia triacanthos*), a tall tree of the same family, but with branched thorns and narrower leaflets. The seed pulp of this latter tree is considered edible, pleasant and sweet (but see also under Kentucky Coffee Tree (p. 79).

This book cannot replace the advice and assistance
of qualified medical personnel.
In all cases of suspected poisoning by plants, or any other substance,
immediate qualified medical advice and assistance should be sought.

Burning Bush and Spindle Tree, or Wahoo
(Staff-tree Family)

Euonymus spp.
(Celastraceae)

Fig. 102. Burning Bush (*Euonymus alatus*).
R. & N. TURNER

QUICK CHECK Shrubs or small trees with perfect 3–5 parted flowers, and lobed fruits surrounded by a bright orange or red covering; leaves bright red or orange in fall; entire plant, especially attractive fruits, potentially dangerous but not known to be fatal.

DESCRIPTION Leafy trees, shrubs or, rarely, vine-like climbers. The branches are usually 4-angled, with conspicuous winter buds. The leaves are simple, opposite, and short-stalked, with smooth or toothed edges. They are mostly deciduous, turning to brilliant red or orange in the fall. Male (pollen-producing) and female (seed-producing) flowers are borne separately. They are small, greenish-yellow or purplish, in small clusters at the leaf axils. The fruits are attractive 3–5 celled capsules, divided to the base into rounded lobes. The seeds, which are white, red, or black, are enclosed in a bright orange or red covering.

OCCURRENCE Although several species of *Euonymus* are native to parts of North America, the one most commonly grown as an ornamental is Burning Bush (*E. alatus*), native to China and Japan. It is a deciduous, spreading shrub up to 2.4 m (8 ft) high. It has branches with characteristic corky wings, and leaves turning bright crimson in the fall.

TOXICITY The European Spindle Tree (*Euonymous europaeus*), and probably all related species, contain a digitalis-like cardioactive glycoside called evomonoside, as well as several peptide and sesquiterpene alkaloids, any of which may be responsible for the toxicity of these species. A protein lectin, inhibiting protein synthesis in cells, has also been isolated. All parts, including leaves, bark, and fruits are poisonous and violently purgative, but human poisoning has been reported only from the fruit of *E. europaeus*. Symptoms of poisoning appear 10–12 hours after ingestion of the fruit (or probably any part of the plant) and include: watery diarrhea, persistent vomiting, fever, chills, weakness, hallucinations, convulsions, and coma.

Fig. 103. European Spindle Tree (*Euonymus europaeus*), fruit. R. & N. TURNER

Fig. 104. Japanese Euonymus (*Euonymus japonicus* 'Silver King'). R. & N. TURNER

TREATMENT Induce vomiting, or perform gastric lavage; follow with activated charcoal and a saline cathartic; supportive treatment for symptoms; immediate replacement of fluids and electrolytes is important.

NOTES Climbing Bittersweet (*Celastrus scandens*) is related to Burning Bush. The attractive fruits of this climbing vine are reputed to be poisonous and should not be eaten. Strawberry Bush (*Euonymus americanus*) is a common shrub of the eastern United States. Although there are no records of human poisoning, it and related native species should be considered dangerous.

Fig. 105. Bittersweet (*Celastrus scandens*), in fruit. MARY W. FERGUSON

Cherry Laurel, or Laurel Cherry
(Rose Family)

Prunus laurocerasus (syn.
  Laurocerasus officinalis)
(Rosaceae)

Fig. 106. Cherry Laurel (*Prunus laurocerasus*), foliage.
R. & N. TURNER

QUICK CHECK Tall shrub, small tree, or hedge plant, with thick, evergreen leaves, whitish flowers in elongated clusters, and blackish, cherry-like fruits. The leaves, bark, fruit, and seeds can cause cyanide poisoning; seldom fatal.

DESCRIPTION Tall evergreen shrub, small tree, or hedge plant, with thick, glossy, leathery leaves. These are lance-shaped to elliptical, short-stalked, pointed, finely toothed around the margins, and up to 15 cm (6 in.) long. The flowers are small and white or cream-colored, in dense, elongated clusters borne in the leaf axils. The fleshy fruits are shiny, dark purple or blackish, each with a single, cherry-like stone in the center. There are many cultivated forms, with different growth habits, leaf sizes and colors.

OCCURRENCE A native of southeastern Europe and the Middle East, this bushy tree or shrub is often planted in North American gardens and parks as an ornamental, shade, and hedge plant.

TOXICITY The leaves, bark, flowers, fruits, and especially the seeds of Cherry Laurel, like those of other *Prunus* species, contain cyanide-producing glycosides, including amygdalin and prunasin. The fully ripe fleshy part of the cherries is probably the least toxic. The symptoms are the same as those of poisoning by Apricot kernels, Apple seeds, and other types of Cherries (see pp. 74, 146), all due to cyanide poisoning. The symptoms may arise very quickly, and death can occur rapidly when large quantities are consumed, but fortunately, this is rare; the human body can handle small quantities without problem. Cherry Laurel poisoning is not common, but children sometimes eat the small, black "cherries," or people sometimes mistake the leaves for those of Bay Laurel (*Laurus nobilis*), and try to use them for flavoring food. The bitter, unpleasant taste is usually a sufficient

Fig. 107. Cherry Laurel, in flower. R. & N. TURNER   Fig. 108. Cherry Laurel, fruits. R. & N. TURNER

deterrent to eating large amounts. Browsing animals are occasionally poisoned by eating the foliage or hedge trimmings of Cherry Laurel.

TREATMENT Induce vomiting, or perform gastric lavage; if symptoms are severe, immediately treat for cyanide poisoning (p. 12); see also under Apricot in this section (p. 146), and Cherries, Wild in Chapter III (p. 74).

NOTES A related species, American Cherry Laurel (*Prunus caroliniana*), is an evergreen tree up to 12 m (40 ft) high with glossy leaves, very small, white flowers in dense clusters, and black, shiny fruit. It is native to moist valleys of the eastern United States from South Carolina to Texas, and is planted as an ornamental in the south-eastern states and southern California.

Cherry Laurel is sometimes confused with the true Laurel, or Bay Laurel (*Laurus nobilis*) in the Laurel Family, whose leaves are used in flavoring soups and stews. Swamp, or Mountain Laurel (*Kalmia* spp.) (p. 80), and Daphne Laurel (*Daphne laureola*) (p. 166) are two other plants which may cause confusion. These are unrelated, but all have leathery, evergreen leaves.

Fig. 109. Chinaberry Tree (*Melia azedarach*).
EUGENE N. ANDERSON

Chinaberry Tree, or China Tree
(Mahogany Family)

Melia azedarach
(Meliaceae)

QUICK CHECK Small to medium deciduous tree with large, divided leaves, small, purplish flowers, and yellowish, cherry-sized fruits; toxins variable in concentration; symptoms may be delayed; sometimes fatal.

DESCRIPTION A deciduous tree up to 15 m (50 ft) tall, with a thick trunk, furrowed bark, and spreading branches. The leaves are large, long-stalked, and compound, each divided into numerous, oval-shaped, toothed leaflets about 2.5–5 cm (1–2 in.) long. The overall leaf shape is roughly triangular. The flowers are lilac-colored, delicate, and fragrant, borne in open, long-stalked clusters. The fruits are fleshy, globular, cream-colored to yellow, and cherry-sized, each containing a single seed. Smooth at first, they persist after the leaves have fallen, becoming wrinkled. There are a number of different varieties; var. *umbraculiformis,* called the Texas Umbrella Tree, has drooping foliage, giving the tree the appearance of a gigantic umbrella; another variety, *floribunda,* has masses of flowers.

OCCURRENCE A native of southwest Asia, this tree is a common garden ornamental and shade plant of the southern United States, particularly from Virginia south to Florida, west to Texas, and in California. It also occurs at lower elevations in Hawaii. It is a frequent garden escape in woods, old fields, fence rows, and scrubby areas.

TOXICITY The leaves, bark, and fruit are toxic, although the concentration of toxins varies from one population to the next. Eating only 6–8 berries can be lethal to a child. The toxins are tetranortriterpene neurotoxins and as yet unidentified resins, causing irritation to the digestive tract and degeneration of the liver and kidneys. Symptoms are often delayed. They include faintness, lack of coordination, confusion, and stupor, and in some cases severe stomach pain, diarrhea, vomiting, difficulty in breathing, convulsions, partial to complete paralysis, and death, which may occur within one day.

TREATMENT Induce vomiting, or perform gastric lavage; follow with activated charcoal and a saline cathartic; fluid and electrolyte replacement as necessary; monitor kidney and liver function.

NOTES This tree has a variety of local names, including Lilac Tree, Bead Tree, China Tree, Paradise Tree, and Pride of India. In some places the fruit is eaten without harm, but its use is definitely not recommended. Children may be poisoned from drinking "tea" made from the leaves. The seeds are sometimes used for rosaries.

Holly, English, and Its Relatives
(Holly Family)

Ilex aquifolium and related
   species
(Aquifoliaceae)

Fig. 110. English Holly (*Ilex aquifolium*), variegated form. R & N. TURNER

QUICK CHECK Dense, evergreen shrub or tree, with dark green, shiny, spiny leaves, small, whitish flowers, and attractive, scarlet berries. Berries and leaves may cause digestive upset; berries an occasional cause of poisoning in children, but not known to be fatal.

DESCRIPTION English Holly is an evergreen shrub or tree, up to 15 m (50 ft) tall, with smooth, gray bark, and many spreading branches forming a dense, conical head. The thick, leathery, dark-green leaves are alternate and short-stalked. The most common varieties have sharp spines at the tip and few to several spine-tipped teeth around a deeply undulating margin. Male and female flowers, borne on separate plants, are small, dull white, and clustered in the leaf axils. The fruits are scarlet, globular, and shiny, each containing 2–4 seeds. Many different varieties are grown, having different sizes, forms, and spine types. Some have variegated leaves, with white or yellowish margins.

OCCURRENCE English Holly is widely grown in temperate North America as an ornamental plant, and for its decorative berries and foliage at Christmas. It is also used as a hedge species. Many other *Ilex* species, both evergreen and deciduous, are found in North America; some are native, others cultivated as ornamentals.

TOXICITY The showy, scarlet berries and the leaves contain glycosides and theobromine, a caffeine-like alkaloid. Eating the berries is the usual cause of poisoning, especially in children. Symptoms, which in young children may occur from eating only a few berries, include nausea, vomiting, diarrhea, and drowsiness. However, fatalities from Holly are unknown, and their poisonous properties are frequently overstated. Mild doses of the leaves or berries cause stimulation of the central nervous system, whereas higher doses cause depression of the central nervous system.

TREATMENT If large quantities of the berries have been injested, induce vomiting (in gastric lavage, berries may block the tube); follow with activated charcoal and a saline cathartic; excess stimulation caused by theobromine can be countered with barbiturates or benzodiazipines. Monitor central nervous system in case of depression caused by high doses of poison.

NOTES Several wild red, orange, and black-fruited Holly species occur in eastern and southern North America. All should be regarded with caution. One native holly, Yaupon Tea (*Ilex vomitoria*), is found along the coast of the southeastern United States. It can be made into a mild tea, but if drunk in a concentrated brew can cause hallucinations and vomiting. It was used by southerners as a substitute for coffee and tea during the American Civil War. Another related beverage plant is the South American Yerba Maté, or Paraguay Tea (*Ilex paraguayensis*), which is still widely used as a stimulating tea. Holly is well known as a Christmas decoration with its bright red berries and dark green, shiny leaves.

Horse Chestnut
(Horse Chestnut Family)

Aesculus hippocastanum
(Hippocastanaceae)

Fig. 111. Horse Chestnut (*Aesculus hippocastanum*), in flower. R. & N. TURNER

QUICK CHECK Large, deciduous tree with palmately compound leaves, whitish or pinkish flowers in upright clusters, and chestnut-like

Fig. 112. Horse Chestnut, leaves and fruits.
R. & N. TURNER

seeds, enclosed in a prickly, greenish husk. The leaves and seeds are toxic, but rarely fatal in humans.

DESCRIPTION A large, much-branching, deciduous tree, up to 30 m (100 ft) high, with smooth, gray bark and prominent, sticky buds in winter and spring. The leaves are opposite and palmately compound, each with 5–7 oblong leaflets radiating from the end of a long stalk. The flowers are borne in dense, often upright, elongated clusters, and are usually white blotched with red and yellow, but there are red- and pink-flowered varieties, as well as double-flowered types. The fruits are spherical capsules, each consisting of a leathery, yellowish green, spiny or warty covering enclosing 1–3 large, conspicuous nut-like seeds. The seeds bear large scars, and are brown and glossy when first exposed, becoming duller with age.

OCCURRENCE Horse Chestnut was introduced from Europe and is widely planted as an ornamental shade tree in North America.

TOXICITY The entire tree, especially the young leaves, sprouts, flowers, and seeds, contains the saponin glycoside aesculin, which breaks down blood proteins. Unidentified alkaloids are also present. Poisoning is most common from children eating the seeds or "conkers" or from people mistaking these for edible Sweet Chestnuts. Animals are sometimes poisoned from eating the leaves and seeds. Symptoms include inflammation of the mucous membranes, vomiting, thirst, weakness, lack of coordination, muscular twitching, dilated pupils, stupor, and paralysis. Coma and death from respiratory paralysis may occur in severe cases, but human fatalities are rare, usually resulting from repeated doses.

TREATMENT Induce vomiting, or perform gastric lavage; administer activated charcoal and a saline cathartic; treat for symptoms; use demulcents to relieve digestive tract irritation; correct fluid and electrolyte balance; monitor respiration and heartbeat.

NOTES Horse Chestnut is a relative of the native Buckeyes, and is similarly toxic (see p. 73). The toxin aesculin is closely related to the hydroxycoumarin found in spoiled Sweet Clover hay (*Melilotus* spp.) from which the anticoagulant rodenticides were originally developed.

The edible Sweet Chestnut (*Castanea* spp.) is unrelated to the Horse Chestnut, and is a popular Christmas and gourmet food.

Laburnum, or Golden Chain
(Bean, or Legume Family)

Laburnum anagyroides (syn.
  L. vulgare)
(Fabaceae, or Leguminosae)

Fig. 113. Laburnum, or Golden Chain (*Laburnum anagyroides*), in flower. R. & N. TURNER

QUICK CHECK Ornamental tree or shrub with bright yellow, pea-like flowers borne in drooping clusters, and hanging, pea-like pods. Pods and seeds are a frequent, not usually fatal, cause of poisoning in children.

DESCRIPTION Large shrub or small tree up to 9 m (30 ft) tall, with spreading branches close to the ground. The leaves, borne on long stalks, are alternate and compound, each with 3 ellipical to oblong leaflets. The flowers are bright yellow and pea-like, in dense, drooping clusters. The fruits are elongated, pea-like pods up to 8 cm (3 in.) long, each containing several seeds, and hanging in clusters. They often overwinter on the branches.

OCCURRENCE Laburnum is native to central and southern Europe. It is grown in gardens throughout temperate North America as an ornamental, for its golden, hanging flowers. Occasionally it occurs as a garden escape.

TOXICITY All parts of the plant, especially the bark and seeds, contain a quinolizidine alkaloid, cytisine, which is similar in its effects to nicotine. The concentration of this potent toxin varies considerably with season and genetic strain. All Laburnum plants should be regarded as highly poisonous, and 20 seeds are reported to be lethal for a small child. However, according to Cooper and Johnson (1984), the reputation of the plant is based on a few dramatic reports of severe poisoning, and most cases are relatively mild.

Poisoning usually results from children eating the green pods or the seeds, mistaking them for peas or beans. The symptoms develop in less than an hour, and include burning of the mouth and throat, nausea, vomiting, abdominal pain, diarrhea, drowsiness, headache, dizziness, fever, rapid or irregular heartbeat, excitment and confusion, sweating, salivation, cold skin, difficult breathing, and dilated pupils. In mild cases, symptoms are restricted to nausea, vomiting,

Fig. 114. Laburnum, showing hanging fruiting pods.
R. & N. TURNER

and mild abdominal pain, and recovery is complete in 12–24 hours. In exceptionally severe cases, hallucinations, convulsions, coma, and sometimes death from circulatory or respiratory failure occur.

TREATMENT Do not induce vomiting nor perform gastric lavage if vomiting has occurred or if seizures are imminent; i.v. diazepam may be used to sedate patient, then administer activated charcoal followed by a saline cathartic (repeat charcoal and cathartic every 6 hours for 24 hours); correct fluid and electrolyte balance; oxygen and artificial respiration may be required; monitor heart beat and arterial blood gases; general supportive treatment for symptoms; control convulsions with i.v. diazepam or phenobarbital; severe constipation and urinary retention at late stages of poisoning respond to bethanechol.

NOTES Despite the bad reputation of this plant, there are few documented fatalities from it. One case involved a man who died without noticeable clinical symptoms, after consuming 23 Laburnum pods; he had apparently absorbed 35–50 mg of the alkaloid cytisine. Laburnum is closely related to Broom (p. 86), and has been placed by some in the same genus, *Cytisus*.

Tung Nut, Candlenut, and Their Relatives
(Spurge Family)

Aleurites fordii, A. moluccana, and related species
(Euphorbiaceae)

Fig. 115. Candlenut (*Aleurites moluccana*).
R. & N. TURNER

QUICK CHECK Medium-sized, deciduous trees with large, heart-shaped or lobed leaves, whitish, clustered flowers, with red or orange veins in Tung Nut, chestnut-like seeds, and milky latex. Entire plants, especially of Tung Nut, toxic and occasionally fatal; the nut-like seeds of Tung Nut are the most common cause of poisoning.

DESCRIPTION There are several species in the genus *Aleurites* found as ornamentals in parts of the southern United States and Hawaii. Two of the most prominent are Tung Nut, or Tung Oil Tree (*A. fordii*) and Candlenut, or Kukui (*A. moluccana*). These are usually medium-sized deciduous trees with large, alternate, smooth-edged leaves that are long-stalked and heart-shaped in Tung Nut, or (usually) shallowly 3–5 lobed in Candlenut. The leaves of Candlenut often have a light, silvery cast. The flowers of these trees occur in large clusters and are usually white, sometimes with red or orange veins in Tung Nut. The fruits, green when unripe and brown at maturity, are large and globular or somewhat pointed, borne on drooping stalks. Within the tough hull, Tung Nut fruits each contain 3–7 rough-coated, chestnut-like seeds; Candlenut seeds are single or paired, with extremely hard shells. All species of the genus should be considered toxic, but Tung Nut is evidently the most poisonous. Candlenut seed kernels are edible when roasted, but toxic when raw.

OCCURRENCE Native to central Asia, Tung Nut is now widely distributed in the tropics. It has been planted in large commercial orchards as an oil crop in the Gulf Coast region of the United States, from northern Florida to Texas, as well as in Hawaii. It is also used as an ornamental and shade tree in the southern States. Candlenut is a native of Polynesia and southeast Asia, and was introduced to the Hawaiian

Islands by the early Polynesian settlers. It is now widespread in the lowland forests of Hawaii and elsewhere in the tropics and sub-tropics. These and other species have been cultivated for the oil pro-duced in their seeds.

TOXICITY All parts of the Tung Nut tree are toxic, containing unidentified saponins, and a phytotoxin—a highly toxic protein molecule, also uncharacterized. It is also reported to contain a derivative of the irritant diterpene ester, phorbol, found in other members of this family. The large, attractive, nut-like seeds are the most common cause of human poisoning; a single seed can cause serious illness, and many cases of poisoning have been reported over the years. Symp-toms include discomfort and nausea, followed by very severe stomach pain, with vomiting, diarrhea, weakness, depressed reflexes and breathing, and sometimes death. Poisoning of cattle and other animals browsing the cut foliage is not uncommon. Candlenut and other *Aleurites* species are also known to be toxic.

TREATMENT Induce vomiting, or perform gastric lavage; treat for symp-toms of severe gastroenteritis, using analgesics if required; replace fluids and correct for electrolyte losses; monitor kidney functions; treat convulsions with parenteral, short-acting barbiturates.

NOTES The spurge family contains a number of toxic and irritant plants, including the Spurges (*Euphorbia* spp., p. 213), Castor Bean (p. 224), the Crotons (p. 227), and at least three other trees found occasionally in the southern United States: Purge Nut, Manchineel Tree, and Sandbox Tree.

Purge Nut (*Jatropha curcas*) is a small tree or coarse shrub with large, alternate, long-stalked, palmately veined leaves that are usually 3–5 lobed, small, yellow flowers, and brownish black fruiting cap-sules containing 2–3 black seeds. It and others in the genus are widely grown as tropical ornamentals, and may be encountered in southern Florida and Hawaii. They contain a toxalbumin, curcin, and the fruits and seeds, attractive to children, cause severe gastrointestinal irrita-tion and sometimes coma when ingested.

Manchineel Tree (*Hippomane mancinella*), native to the southern tip of Florida and the Keys, but now restricted to remote areas like the Everglades, is a small to medium-sized tree with milky juice, large, oval, dark green leaves with finely toothed margins, small, greenish flowers in spikes, and round, green or yellowish green fruits about 4 cm (1.5 in.) across. The milky sap is extremely caustic to the skin and can cause temporary blindness, and the seeds cause severe gastroen-teritis if ingested.

Sandbox Tree (*Hura crepitans*) is a large, spiny tree occasionally grown for its curious, violently explosive fruits. Its seeds and milky juice are highly irritant, causing severe vomiting and diarrhea, and are also carcinogenic, due to the presence of a diterpene, huratoxin.

Yew, or Ground Hemlock
(Yew Family)

Taxus spp.
(Taxaceae)

Fig. 116. Yew (*Taxus baccata*), showing fruits.
R. & N. TURNER

QUICK CHECK Evergreen, needled trees or shrubs with pollen cones and seeds borne on separate plants; seeds surrounded by a fleshy, reddish cup; entire plant, except fleshy "berry" around the seed, is highly toxic, but human fatalities are rare.

DESCRIPTION Evergreen trees or shrubs with scaly, reddish-brown bark. The branches have a flattened appearance, with the flat, needle-like leaves spreading out along the twigs. The needles are pointed at the tip, dark green above, and pale green beneath. The male (pollen-producing) cones and the female (seed-producing) structures are borne in early spring and occur on separate plants. Each ripened seed is surrounded by a pinkish to scarlet, fleshy cup, or aril, giving it the appearance of a "berry" with the seed exposed at the top.

OCCURRENCE English Yew (*Taxus baccata*) and Japanese Yew (*T. cuspidata*) are the most commonly grown ornamental Yews in North America, and are found in gardens and hedges, lending themselves well to ornamental shaping. There are three species native to North America and growing in forests and woodlands: Western Yew (*T. brevifolia*), in western North America from British Columbia to California; Ground Hemlock (*T. canadensis*) in the East from Newfoundland to Virginia and the midwestern states, and Florida Yew (*T. floridana*), restricted to the Apalachicola River area in Florida.

TOXICITY All parts of Yews, except the fleshy red aril around the seed, contain significant amounts of taxine, a complex mixture of alkaloids absorbed rapidly by the human digestive system and acting on the heart. Yews also contain another alkaloid, ephedrine, as well as an irritating volatile oil, and a cyanogenic glycoside, taxiphyllin. Symptoms of Yew poisoning include nausea, dry throat, severe vomiting, diarrhea, rash, pallor, drowsiness, abdominal pain, dizziness, trembling, stiffness, fever, and sometimes allergy symptoms. Occasional fatalities occur when large quantities are consumed (as when a women in Germany reportedly ate 4–5 handfuls of the leaves), or when children eat the "berries," chewing up the seeds. In the last case, even one or two seeds could be lethal for a small child. Abdominal pain, irregular heart beat, dilated pupils, collapse, coma, and convulsions, followed by slow pulse and weak breathing, are symptoms of severe poisoning. Death is from respiratory and heart failure. Parents of small children should remove the decorative "berries" from any nearby ornamental yews. Browsing animals are frequently fatally poisoned by Yew.

TREATMENT Induce vomiting, or perform gastric lavage; follow with activated charcoal and a saline cathartic; supportive treatment for symptoms; correct fluid and electrolyte balance; monitor breathing, heart beat, and blood pressure; artificial respiration and use of pacemaker may be required. Dopamine may be used to treat persistent low blood pressure, and i.v. diazepam for convulsions.

NOTES Yews have been used since prehistoric times in folk medicine, both in Europe and North America. Occasional deaths have occurred from misuse of herbal preparations from the leaves. The fleshy red aril, or "berry," is pleasant tasting and edible, at least in small quantities and as long as the seed is not eaten with it. Even then, there have been cases of as many as 40 berries being eaten, with the seeds, without any symptoms of poisoning being evident, apparently because the seeds were swallowed without being chewed and passed through the digestive tract without releasing their toxins.

This book cannot replace the advice and assistance
of qualified medical personnel.
In all cases of suspected poisoning by plants, or any other substance,
immediate qualified medical advice and assistance should be sought.

MEDIUM-TO-LOW SHRUBS AND HEDGE PLANTS

Aucuba
(Dogwood Family)

Aucuba japonica
(Cornaceae)

Fig. 117. Aucuba (*Aucuba japonica*). R. & N. TURNER

QUICK CHECK Evergreen shrub with large, coarsely toothed, often mottled leaves, and scarlet, berry-like fruits; fruit mildly poisonous.

DESCRIPTION A medium-sized to large evergreen bush with opposite leaves, in some varieties mottled with yellow spots or variously marked. The leaves are up to 18 cm (7 in.) long, elliptical, and coarsely toothed along the margins. The flowers are purple, borne in clusters at the ends of the branches. The berry-like fruits are scarlet and some-what elongated, containing a single seed.

OCCURRENCE Native to Asia, this shrub is often cultivated as an ornamental along the Atlantic and Pacific coasts.

TOXICITY The entire plant contains an acid-modifying glycoside, aucubin. So far, however, symptoms of poisoning have only been reported from eating the fruit, and even then they are not always apparent. They may include digestive upset, vomiting, and fever.

TREATMENT Treat for symptoms; if vomiting is persistent, correct fluid and electrolyte balance.

NOTES This plant is somewhat similar in appearance to the cyanide-producing Cherry Laurel (p. 152), but is easily distinguishable in the fruiting stage by its larger, scarlet fruits borne in clusters that are not elongated.

Boxwood, or Common Box
(Box Family)

Buxus sempervirens
(Buxaceae)

Fig. 118. Boxwood (*Buxus sempervirens*). R. & N. TURNER

QUICK CHECK Evergreen shrub with simple, leathery leaves, commonly grown as a hedge plant; potentially fatal in large quantities.

DESCRIPTION Evergreen shrub or small tree (rarely) up to 8 m (25 ft) high, with 4-angled or slightly winged branches. The leaves are short-stalked, opposite, and leathery, up to 4 cm (1.5 in.) long. The flowers, formed in early spring, are small, pale green, and inconspicuous. The fruit is a small, ovoid capsule. There are many horticultural varieties, differing in growth form, leaf size and color; some have leaves that are variegated or edged with yellow or silvery white. A related species, Japanese Box (*B. microphylla,* or *B. japonica*), is also grown in North America. It is a compact low or prostrate shrub, usually with smaller leaves than Common Box.

OCCURRENCE A native of western Europe and the Mediterranean, Box is extensively used in North America for hedges and borders because it stands pruning and shaping well. It is also sometimes grown as a single ornamental shrub or small tree.

TOXICITY The leaves and twigs contain a complex group of steroidal alkaloids, known collectively as buxine. Eating them can cause abdominal pains, vomiting, diarrhea, and, in large doses, lack of coordination, convulsions, coma, and death from respiratory failure. Fortunately, the plant is not particularly conspicuous, and seldom causes poisoning in humans. Browsing animals, however, have been fatally poisoned from Box hedge clippings, and in one instance, aquarium fish were killed when the branches were placed in the tank as decoration.

TREATMENT Induce vomiting, or perform gastric lavage; follow with activated charcoal and a saline cathartic; treat for symptoms.

NOTES Although quite toxic, Box is usually avoided by animals because of its disagreeable odor and acrid taste. It is sometimes confused with a low, evergreen shrub native to western North America called False Box (*Paxistima myrsinites*). This latter shrub, often used in floral decorations, is related to the poisonous Burning Bush and Spindle Tree (*Euonymus* spp.), but has not itself been implicated as poisonous.

Daphne, or Mezereon
(Mezereum Family)

Daphne mezereum and
related species
(Thymelaeaceae)

Fig. 119. Daphne (*Daphne mezereum*), berries.
R. & N. TURNER

QUICK CHECK Deciduous shrub with simple, alternate leaves; purple or white, fragrant flowers appear in spring before the leaves; scarlet, or occasionally yellow, single-seeded, berry-like fruits. The entire plant, especially the attractive fruits, is highly toxic; the fruits may be fatal if only a few are eaten.

DESCRIPTION A deciduous shrub up to 1.2 m (4 ft) high, with simple, alternate, bright green leaves 5–8 cm (2–3 in.) long. The flowers are small fragrant, lilac-purple, or sometimes white, growing in small clusters and appearing in spring before the leaves. They are funnel-shaped and 4-lobed, flaring at the mouth. Fruits are berry-like drupes, scarlet, or yellow in the white-flowered variety, and very attractive. A related shrub, Laurel Daphne, or Spurge Laurel (*Daphne laureola*), has dark green, leathery, evergreen leaves, yellowish-green, non-scented flowers, and bluish black fruits. Another evergreen Daphne, having bright red berries, is *D. retusum*. Both are also poisonous.

OCCURRENCE A native of Eurasia, Daphne is cultivated as an ornamental throughout much of North America, and occurs as a garden escape in the Northeastern States and eastern Canada. Laurel Daphne and other *Daphne* species are also widely grown as ornamentals, and are sometimes found as escapes.

TOXICITY All parts of the plant, particularly the bark and berry-like fruits, contain an acrid, irritant, highly toxic sap. Several poisonous compounds are present, including: a dihydroxycoumarin glycoside, daphnin, and its hydrolysis product, daphnetin; a hydroxycoumarin, umbelliferone; a resin, mezerein, or its derivatives, and a diterpene alcohol, daphnetoxin. Some of these components are referred to generally in the literature as glycosides, and daphnin as a lactone glycoside. The coumarins are believed to be responsible for the acrid taste, whereas mezerein is considered to be the most toxic component.

Chewing the bark or fruits causes painful blistering of the lips, mouth, and throat, with salivation, thirst, and inability to eat or drink, followed by swelling of the eyelids and nostrils, intense burning and ulceration of the digestive tract, vomiting, bloody diarrhea, weakness, headaches, and in severe cases, delirium, convulsions, coma, and death. Only a few fruits may be fatal to a child. The sap of Daphne may cause severe skin irritation and ulceration, and the poison may enter the body through skin contact.

TREATMENT Induce vomiting, or perform gastric lavage; follow with activated charcoal and a saline cathartic; use demulcents such as milk or ice cream to coat and soothe the mouth and digestive tract, or allow patient to suck on ice chips or a popsicle; monitor kidney function; maintain fluid and electrolyte balance; treat for symptoms; antacids may provide relief; avoid antidiarrheal drugs; treat for shock if present; symptoms may persist for several days.

NOTES Reference to Daphne as a poisonous plant is found in the writings of early Greek herbalist Dioscorides. Cases of Daphne poisoning in North America are rare, but the brightly colored fruits may attract young children, and since only a few berries are enough to kill a child, people growing Daphne in conspicuous places should take care to remove the berries in order to protect neighbourhood children.

In Victoria in the summer of 1989, a young child ate a few of the bluish black berries of Daphne Laurel (*D. laureola*). He exhibited typical symptoms of Daphne poisoning, but recovered after treatment.

According to Levy and Primack (1984), some extracts of Daphne have shown potential in treating cancer, notably leukemia.

Fig. 120. Laurel Daphne (*Daphne laureola*), in flower.
R. & N. TURNER

Hydrangeas
(Hydrangea Family)

Hydrangea spp.
(Hydrangeaceae)

Fig. 121. Hydrangea (*Hydrangea macrophylla*), cultivar. R. & N. TURNER

QUICK CHECK Broad-leaved shrubs with opposite, coarsely toothed leaves and flowers in dense rounded or flat clusters. Leaves and buds of some species contain a cyanide-producing compound, causing illness, but no reported fatalities.

DESCRIPTION There are at least four species of *Hydrangea* native or grown as ornamentals in North America. They are low to tall shrubs with opposite, simple, elliptical or oval leaves that are stalked, coarsely toothed, or sometimes lobed. The flowers grow in rounded to flat-topped clusters up to 15 cm (6 in.) or more across, and range in color from white to pink to blue. The flowers themselves are sterile, with the showy colored part being an expanded calyx.

OCCURRENCE Some Hydrangeas are native to eastern North America. The major ornamental species, *H. macrophylla,* a native of Japan, is now found in gardens throughout the temperate region.

TOXICITY The leaves and buds of some *Hydrangea* species, but apparently not of one of the most commonly cultivated types, *H. paniculata* 'Grandiflora', contain a cyanide-producing glycoside, hydrangin. *H. paniculata* contains other harmful compounds and has caused illness to people smoking the leaves in search of a marijuana-like "high." Symptoms of Hydrangea poisoning, from eating the leaves and buds, or from smoking the leaves, include nausea, vomiting, and diarrhea. Apparently no fatalities have been reported.

TREATMENT Induce vomiting, or perform gastric lavage; be prepared to treat for cyanide poisoning (p. 12).

NOTES The rhizome and roots of the Arborescent Hydrangea (*H. arborescens*), native to the eastern United States, were originally used by the Cherokee Indians and the early settlers as a remedy for kidney stones. Later, the drug, called "seven barks," was used commercially as a diuretic, but its efficacy is doubtful and its use not recommended. Similarly, thrill-seekers are warned against smoking the leaves of the common garden Hydrangea.

Pieris
(Heather Family)

Pieris japonica
(Ericaceae)

Fig. 122. Pieris (*Pieris japonica*). R. & N. TURNER

QUICK CHECK Low to tall shrub with simple, leathery leaves, often reddish when young, and dense sprays of white (or pinkish), urn-shaped flowers; occasionally fatal to children.

DESCRIPTION Shrub or small tree, rarely as high as 10 m (30 ft). The leaves are simple, leathery, smooth, toothed, and up to 8 cm (3 in.) long. The young leaves are often scarlet or pinkish. The small, urn-shaped flowers are white or pinkish, clustered in dense, showy sprays. The fruits are small 5-parted capsules.

OCCURRENCE This shrub is a native to Japan. Many forms are planted as ornamentals in North American gardens, including a dwarf "bonsai" form, and one with variegated leaves. There are several other species, including two native to North America. All should be considered dangerous.

TOXICITY Like many other plants in the heather family, Pieris contains toxic diterpenoids called grayanotoxins (including andromedotoxins). The leaves, and even honey made from the flower nectar, are toxic. Symptoms are: burning sensation in the mouth, followed by salivation, copious tears, runny nose, vomiting, stomach cramps, diarrhea, and "prickly" skin after several hours. Headache, dim vision, weakness, and slow heartbeat, followed by severe hypotension, coma, and convulsions may also occur. Fatalities in children eating the leaves have been reported.

TREATMENT Induce vomiting, or perform gastric lavage; follow treatment for Rhododendrons and Azaleas (p. 171).

NOTES Pieris, also known as Lily-of-the-Valley Bush, is related to several other poisonous shrubs with similar toxins and symptoms, including Rhododendrons and Azaleas (p. 171), and Mountain Laurel and its relatives (p. 80).

Fig. 123. Privet (*Ligustrum vulgare*), flowers.
R. & N. TURNER

Privet
(Olive Family)

Ligustrum vulgare
(Oleaceae)

QUICK CHECK Deciduous (sometimes evergreen) shrub and hedge plant with medium-sized leaves, small, white, clustered flowers, and small, shiny black berries. Leaves and berries potentially, but rarely, fatal.

DESCRIPTION Densely growing hedge plant or open shrub up to 5 m (16 ft) tall, with smooth bark, slender branches, and young shoots having a light covering of hairs. The more trimming the bushes receive, the denser their branches. The leaves, usually 3–6 cm (1–2.5 in.) long, dark green above, lighter beneath, are borne in pairs. They are simple and elliptical, with short stalks and smooth edges. Although classed as deciduous, the plant is partially evergreen in some varieties. The flowers are small, creamy-white, tubular, and fragrant, in dense pyramidal clusters; flowering is in summer. The tightly clustered berries are dark blue or black, and shiny, the size of small peas. They may remain on the bushes over the winter, making them con-spicuous to children after the leaves have fallen.

OCCURRENCE Native to southern England, Privet is one of the most widely cultivated hedge plants in North America, and is also grown singly as an ornamental shrub. It is sometimes found as a garden escape in woodlands near settled areas. There are many cultivated varieties, with variegated leaves, small leaves, or dwarf in size.

TOXICITY The main poisonous compound, found throughout the plant and especially in the berries, is an irritant glycoside, ligustrin or syringin. Human poisoning is rare, but has occurred from children eating the berries and adults chewing on the leaves. Symptoms, including severe gastrointestinal irritation, with vomiting and diarrhea, may persist from 48 to 72 hours. Occasional fatalities, preceded by convulsions, have been reported. Animal poisoning, usually from eating the hedge trimmings, is unusual, but when it does

Fig. 124. Privet, berries.
R. & N. TURNER

occur, is often severe or even fatal. Privet may also cause severe dermatitis in people trimming hedges or bushes.

TREATMENT Induce vomiting, or perform gastric lavage; follow with activated charcoal and a saline cathartic; replace fluids to prevent dehydration; treat for symptoms.

NOTES Another species, Oval-leaved Privet (*Ligustrum ovalifolium*), is also commonly grown as an ornamental and hedge plant.

Rhododendrons and Azaleas
(Heather Family)

Rhododendron spp.
(Ericaceae)

Fig. 125. Rhododendron (*Rhododendron* hybrid).
R. & N. TURNER

QUICK CHECK Evergreens (mostly Rhododendrons), or deciduous (mostly Azaleas) shrubs with simple leaves and large, showy, bell-shaped or funnel-formed flowers of various colors; leaves, flowers, and flower nectar toxic, but rarely fatal for humans.

DESCRIPTION There are about 800 species of *Rhododendron,* and innumerable cultivated varieties. They are evergreen or deciduous shrubs, or rarely small trees, with prominent, tightly scaled buds. They are generally divided into two groups—Rhododendrons, most with leathery, evergreen leaves, and Azaleas, most with thinner, deciduous leaves. The leaves of both are short stalked, simple, and alternate, and the showy flowers are large, bell-shaped or funnel-formed, and borne in dense clusters. They come in many colors: white, cream, pink, mauve, red, orange, yellow, and purple. The fruits are dry capsules containing numerous tiny seeds.

OCCURRENCE *Rhododendron* species occur naturally in temperate regions of the northern Hemisphere and in the mountains of Southeast Asia and Australia. There are 27 species native to North America. Rhododendrons and Azaleas are extremely popular as ornamentals in North American gardens, and the hybridization of various species has led to a major horticultural industry.

TOXICITY The leaves, flowers, pollen, and nectar of many Rhododendron species contain several toxic diterpenoids called grayanotoxins. One of these, grayanotoxin I, or rhodotoxin (sometimes also called andromedotoxin, but this name has also been used for another toxin), is prevalent in Rhododendron flower nectar, and has caused poisoning of bees and the honey they produce from it. Human poisoning from eating the plant itself is unusual, but has been reported in children who chewed the leaves. Symptoms of poisoning include initial burning of the mouth, salivation, watering eyes, and runny nose, followed up to several hours later by vomiting, convulsions, headache, "prickly skin," muscular weakness, drowsiness, slow and irregular heart beat, very low blood pressure, and paralysis. Coma and convulsions, followed by death, may occur in extreme cases. Chronic poisoning, from drinking tea of Rhododendrons or their relatives, or using Rhododendron honey, may occur, with persistent low blood pressure and episodes of high blood pressure.

Animal poisoning and fatalities, from browsing the leaves or clippings, have been frequent, usually occurring during times of food shortage when edible species are scarce. Symptoms are similar to those of human poisoning.

TREATMENT Induce vomiting, or perform gastric lavage; follow with activated charcoal and a saline cathartic; replace fluids and maintain electrolyte balance; monitor heart beat, blood pressure, and breathing; if required, give atropine to increase heart rate, and ephedrine or dopamine for low blood pressure; be alert for episodes of high blood pressure; artificial respiration and oxygen may be needed; convulsions and restlessness may be controlled with i.v. diazepam. Recovery should be complete in 24 hours.

NOTES As early as 400 B.C. a report was made of Greek soldiers being poisoned by wild Rhododendron honey. Bees may also be affected; in parts of Scotland, ornamental Rhododendron plantings have made beekeeping uneconomic because of serious loss of bees in the spring from Rhododendron nectar.

VINES

Clematis, or Virgin's Bower
(Buttercup Family)

Clematis spp.
(Ranunculaceae)

Fig. 126. Clematis (*Clematis montana*). R. & N. TURNER

QUICK CHECK Trailing or climbing woody vines (occasionally erect), with usually compound leaves and showy flowers of various colors; the plants contain a severe irritant; seldom fatal.

DESCRIPTION Most Clematis are woody vines, climbing by clasping leaf-stalks. There are also some erect, perennial herbs. The leaves are opposite, and usually divided into three or more segments. The flowers range from small to large and very showy, with colors from white to pink to reddish to purple. Some are urn-shaped or bell-shaped; most open flat at maturity, revealing numerous stamens. The colored portions are actually sepals; petals are lacking. In some cases, the flowers are clustered, but most ornamental types have large, singly borne flowers. The densely clustered fruits are small and single-seeded, each with a long, feathery style.

OCCURRENCE There are over 200 *Clematis* species, many native to temperate North America. Species grow wild in many regions of the continent, but the plants are usually encountered as cultivated ornamental vines on walls, fences, and trellises. Many horticultural varieties are available, with large, attractive flowers in a wide range of colors.

TOXICITY Clematis and many other plants in the buttercup family contain an irritant compound, protoanemonin, in the fresh leaves and sap. Produced from a glycoside, ranunculin, it can cause irritation and blistering of the skin with handling of the plant, and, if the leaves are eaten, it causes intense burning and inflammation of the mouth and digestive tract, with profuse salivation, inflamed eyes, blistering, and ulceration, followed by abdominal cramps, weakness, vomiting of blood, and bloody diarrhea. The kidneys may also be affected; initially, there may be excessive, painful, urination with the presence of blood in the urine, followed by a depletion in urine output. The victim may also feel dizzy and confused, and in severe cases, suffer fainting and convulsions. Fortunately, the initial burning of the mouth usually precludes eating large, or fatal doses. Animals are occasionally poisoned from eating Clematis, but because of the acrid taste and irritation of the mouth caused by the plant, fatalities are not common.

TREATMENT If a substantial quantity has been eaten, induce vomiting, or perform gastric lavage; administer demulcents to sooth irritated membranes; correct fluid balance; urine output should be monitored.

NOTES A further discussion of protoanemonin and its effects is given under Buttercups (*Ranunculus* spp.) in Chapter III (p. 103). The compound is present only in fresh plant material; drying or cooking apparently deactivates it.

Ivy, English
(Ginseng Family)

Hedera helix
(Araliaceae)

Fig. 127. English Ivy (*Hedera helix*). R. & N. TURNER

QUICK CHECK Evergreen vine of gardens and woods, with dark green or mottled, leathery leaves, usually 3–5 lobed; flowers small, greenish, clustered; fruits pea-sized and black. Leaves and berries moderately poisonous, seldom fatal. Keep small children away from berries.

DESCRIPTION Creeping or climbing evergreen vine with many short, clinging aerial roots along the stems. Leaves leathery, pungent smelling when crushed, usually dark green and 3–5 lobed, but there are horticultural forms with leaf form varying from simple shield shaped to long-pointed star shaped, and leaf color from dark green to variegated white, cream, or yellow. The leaves range in size from 4–10 cm (1.5–4 in.) across. The flowers are whitish to greenish and small, in rounded clusters, with male and female flowers on separate plants. The fruits are round, black, bitter-tasting, pea-sized berries, each containing 3–5 seeds.

OCCURRENCE Native to Europe and the United Kingdom, this Ivy is widely planted as a ground and wall cover throughout North America. A common garden escape, growing in patches on the ground and climbing trees in woods near settled areas. Many horticultural varieties are grown as houseplants.

TOXICITY The entire plant contains triterpene saponins, called hederasaponins A and B, which undergo partial hydrolysis, with loss of sugars, to form toxic saponin glycosides, hederins. The berries occasionally poison children who eat them in quantity. Reported symptoms include vomiting, diarrhea, laboured breathing, excitation, convulsions, and coma, but there have been few recent cases with such drastic results. The leaves are harmful to browsing livestock if large amounts are eaten, with symptoms of vomiting, diarrhea, spasms, staggering, and paralysis; animals usually recover after a few days. The sap can cause dermatitis, sometimes with severe blistering and inflammation.

TREATMENT If a child has eaten a large number of berries, induce vomiting, or perform gastric lavage; follow with activated charcoal and a saline cathartic; treat for symptoms; in severe cases, use paraldehyde (2–10 cc) IM; oxygen and artificial respiration as necessary.

NOTES A related plant, Hercules' Club, or Devil's Walkingstick (*Aralia spinosa*), of the eastern United States, has black berries that may also be poisonous. The red-colored berries of another relative, Devil's Club (*Oplopanax horridus*), are also reputedly poisonous. This latter plant is widely used in Native Indian medicine. All are related to the Ginseng, a well-known Chinese herbal medicine and tonic.

Fig. 128. English Ivy, in fruit. R. & N. TURNER

Fig. 129. Variegated Ivy (*Hedera helix* 'Glacier'). R. & N. TURNER

Jessamine, Yellow, or Carolina
(Logania Family)

Gelsemium sempervirens
(Loganiaceae)

Fig. 130. Yellow Jessamine (*Gelsemium sempervirens*).
WALTER H. LEWIS

QUICK CHECK Trailing or climbing woody vine with yellow, aromatic tubular flowers in early spring; entire plant poisonous, especially flower nectar; fatalities rare.

DESCRIPTION A perennial, evergreen, woody vine, trailing or climbing on bushes and trees. The leaves are paired, short-stalked, and lance-shaped, up to 10 cm (4 in.) long, with smooth margins. The flowers are bright yellow and funnel-form, appearing in early spring and usually clustered. They are showy, up to 4 cm (1.5 in.) long, and very aromatic. The fruits are thin, flattened capsules with winged seeds.

OCCURRENCE Yellow Jessamine is common along the southeastern seaboard, north to eastern Virginia and west to Texas. It is found in woodlands, fields, fencerows, and thickets, usually climbing on trees or fences. It is said to be one of the most beautiful vines of the early spring in the southeast, and is the state flower of South Carolina. It is commonly cultivated throughout its natural range and in southern California.

TOXICITY The entire plant contains potent alkaloids, including gelsemine, gelseminine, and gelsemoidine, with greatest concentrations in the roots and flower nectar. Poisonings may result from misuse of the plant as an herbal medicine. Children have been poisoned by chewing on the leaves and sucking the nectar. Symptoms of poisoning include headache, dizziness, visual disturbances, profuse sweating, and in severe cases, muscular weakness, convulsions, muscular spasms, depression, paralysis of the motor nerve endings, and sometimes death through respiratory failure. Honeybees have been poisoned by the flower nectar, and honey made from the nectar may also be toxic. The plant is also considered toxic for all types of livestock.

TREATMENT Induce vomiting, or perform gastric lavage; administer activated charcoal and a saline cathartic; treat for symptoms; atropine (2 mg. IM) has been suggested; monitor breathing; artificial respiration may be required.

NOTES The well-known poison, strychnine, is produced by a plant in the same family as Yellow Jessamine. This Jessamine may be confused with another plant, also sometimes called Jessamine (*Cestrum* spp.), in the nightshade family. At least three species of *Cestrum* are grown as ornamentals in the southern United States, and may also occur in the wild. All parts of these plants are poisonous, with symptoms similar to those of poisoning by their relatives, Belladonna (*Atropa belladonna*) (p. 183) and Jimsonweed (*Datura stramonium*) (p. 118). DO NOT administer atropine for *Cestrum* poisoning.

Morning Glory
(Morning Glory Family)

Ipomoea tricolor (syn. *I. violacea*)
(Convolvulaceae)

Fig. 131. Morning Glory (*Ipomoea tricolor*).
R. & N. TURNER

QUICK CHECK Annual or perennial vine with large, showy, flaring flowers. Seeds hallucinogenic, potentially dangerous to children, but not fatal.

DESCRIPTION Vigorous annual perennial vine without hairs on the stem or leaves. The twining stems grow up to 2 m (6 ft) high. The leaves are heart-shaped, usually from 10–25 cm (4–10 in.) long. The funnel-shaped flowers are large and showy, up to 10 cm (4 in.) across, varying in color from pale to sky blue, blue-and-white striped, pink, and rose (all with paler centers), or white, depending on the variety. Each flower lasts only one day. The fruiting capsules contain numerous seeds.

OCCURRENCE Morning Glory varieties are often grown in North America as ornamental vines. Additionally, several cultivars, including 'Heavenly Blue', 'Pearly Gates', 'Flying Saucers', 'Wedding Bells', 'Summer Skies', and 'Blue Star', have attained popularity because of their hallucinogenic effects.

TOXICITY The seeds of various cultivated varieties of the species *I. violacea*, but not necessarily of other species of the genus *Ipomoea*, contain amides of lysergic acid, which are similar to but not nearly as potent as LSD in their hallucinogenic effects. They are sometimes eaten or drunk as an infusion by those seeking a "high." Ingestion of the seeds can cause nausea and hallucinations. Their main danger is when small children ingest relatively large quantities, which is, fortunately, not a common occurrence. Additionally, those wishing to experiment with these seeds should be warned that some major suppliers dust them with a noxious chemical fungicide.

TREATMENT General supportive therapy for symptoms; if necessary, treat for acute psychotic reaction, as for LSD treatment, but effects not as severe; keep patient quiet and provide reassurance.

NOTES The seeds of the Morning Glory vine and its relatives have been used for centuries in Central America and elsewhere for their hallucingenic effects. Characteristic visions of the "little people," common to those who use Morning Glory seeds or any of a number of other hallucinogens, may well account for the worldwide traditions of elves, leprechauns, gnomes, and other tiny people of folk tales.

Virginia Creeper, American Ivy, or Woodbine
(Grape Family)

Parthenocissus quinquefolia
(Vitaceae)

Fig. 132. Virginia Creeper (*Parthenocissus quinquefolia*), leaves and fruits. R. & N. TURNER

QUICK CHECK Woody vine with palmately divided, compound leaves and small, blue or blackish, clustered fruits; berries toxic, suspected of causing fatalities in children.

DESCRIPTION A woody vine, climbing by many-branched tendrils, with long-stalked, alternate, compound leaves divided into 3–6 radiating leaflets, each pointed and toothed along the margins. The flowers are small, greenish, and clustered, and the fruits are small, blue or blackish, grape-like berries. A closely related species, *Parthenocissus vitacea*, can be distinguished by having few-branched tendrils. There are several cultivated varieties of Virginia Creeper.

OCCURRENCE Native to woods of eastern and central North America, Virginia Creeper is commonly cultivated for its decorative foliage, which turns scarlet in the fall. It grows on walls, fences, and trees throughout much of the United States and southern Canada.

TOXICITY The berries have been implicated in several cases of poisoning, and at least some fatalities, in children eating them. The evidence has been circumstantial, and the actual toxins are apparently unknown. In one study, however, a guinea pig fed 12 berries was fatally poisoned within 36 hours. The plants are known to contain irritant needle-like crystals, or raphides, similar to those of members of the arum family. Symptoms of poisoning include vomiting, diarrhea, and feeling a need for urination or a bowel movement.

TREATMENT Induce vomiting, or perform gastric lavage; treat for symptoms.

NOTES This vine is related to Grape, whose fruits and leaves are edible. A close relative of Virginia Creeper, Boston or Japanese Ivy (*P. tricuspidata*), is also potentially toxic and contains irritant rhaphides. It has simple, glossy, 3-lobed, (sometimes 3-parted) leaves, and is commonly grown as an ornamental. Virginia Creeper is sometimes confused with Poison Ivy (p. 87), but the latter has three leaflets per leaf.

Fig. 133. Boston Ivy (*Parthenocissus tricuspidata*).
R. & N. TURNER

Wisteria
(Bean, or Legume Family)

Wisteria spp.
(Fabaceae, or Leguminosae)

Fig. 134. Wisteria (*Wisteria sinensis*). R. & N. TURNER

QUICK CHECK Woody vines with pinnate leaves and purplish to white
   pea-like flowers in hanging clusters; entire plant, especially seeds and
   seedpods, toxic, but seldom fatal.

DESCRIPTION Woody, deciduous climbing vines, or occasionally grown as
   shrubs or trees. The leaves are pinnately compound, and the fragrant,
   pea-like flowers are produced in large, showy, hanging clusters. They
   are usually purplish or blueish, but white-and pink-flowered varieties also exist. The hanging, flattened fruiting pods are bean-like and
   smooth or velvety, depending on the species.

OCCURRENCE There are several Wisteria species native to North America,
   growing in rich woods and around wooded swamps of the eastern
   United States. The major cultivated types, including *W. sinensis* and *W.
   floribunda,* are introduced from Asia, and are commonly planted as
   twining ornamentals around patios, trellises, and walls, or sometimes grown as bushy specimens on lawns. The cultivated species
   also commonly occur as garden escapes, especially in the southeastern United States.

TOXICITY All parts of the plant, especially the seeds and seed pods, are
   poisonous. The toxin is composed of an unidentified glycoside, wis-
   tarin, and a lectin, or a poisonous resin. Eating one or two seeds,
   which resemble lima beans, can cause serious illness in a child. The
   toxin is an irritant to the digestive system, with symptoms including
   nausea, abdominal pains, and repeated vomiting, sometimes with
   mild diarrhea. Ingesting large amounts of the bark has caused
   hypovolemic shock from fluid loss. Even in severe cases recovery is
   usually complete in 24 hours.

TREATMENT Induce vomiting, or perform gastric lavage initially; however, since poisoning symptoms include repeated vomiting,
   antiemetics and fluid replacement may be required; supportive treatment for symptoms.

NOTES It is reported in *Sturtevant's Edible Plants of the World* (Hedrick 1972)
   that Wisteria flowers were eaten in China. They should, however, be
   considered as toxic as the rest of the plant until proven otherwise.

FLOWERS AND HERBACEOUS PLANTS

**Autumn Crocus, or
Meadow Saffron**
(Lily Family)

Colchicum autumnale
(Liliaceae)

Fig. 135. Autumn Crocus (*Colchicum autumnale*), spring foliage. R. & N. TURNER

QUICK CHECK Autumn-flowering corm plants with long, tubular mauve to purplish, showy flowers; entire plant highly toxic.

DESCRIPTION A perennial growing from a corm up to 5 cm (2 in.) across. The leaves are glossy and bright green, up to 30 cm (12 in.) long, and 2.5 cm (1 in.) or more wide. They are produced in the spring, then die back before the flowers emerge, in the late summer and fall. The flowers, usually produced in quantity from many, clustered corms, are showy, long, tubular at the base, and pale purple (or sometimes almost white). The flowers die down soon after opening, and an oval fruit is formed at the base of the flower tube, close to the ground, where it remains until the following spring. The fruiting stalk then elongates and the ripened capsules appear above the leaves, splitting open to reveal many small seeds.

OCCURRENCE Commonly grown in gardens and as a potted bulb flower indoors.

TOXICITY All parts of the plant, particularly the corm and the seeds, are toxic, containing the alkaloid drug colchicine and a related alkaloid, colchiceine. Colchicine affects the nervous system and prevents normal cell division. Symptoms of poisoning include burning pain in the mouth and throat, intense thirst, nausea, and violent vomiting. Severe abdominal pain and profuse, persistent diarrhea, followed by lethargy, low blood pressure, shock due to fluid loss, and collapse, have been reported. Kidney damage may occur, with scanty and blood-stained urine. Convulsions, coma, and death from respiratory

Fig. 136. Autumn Crocus, flowers. R & N. TURNER

failure may occur. Colchicine is slow to be absorbed, and signs of poisoning may be delayed for as long as 12–48 hours. Temporary hair loss may occur after about two weeks. All classes of animals are susceptible to colchicine poisoning. A 16-year old was fatally poisoned from eating more than 12 Autumn Crocus flowers. Children and young animals may be poisoned from colchicine-contaminated milk.

TREATMENT Induce vomiting or perform gastric lavage, followed with activated charcoal and a saline cathartic; maintain fluid and electrolyte balance; monitor blood pressure, urine output, serum potassium, respiration, and heart rate; analgesics to relieve the abdominal pain, and atropine to alleviate the diarrhea are suggested. Artificial respiration may be necessary. Symptoms are usually long-lasting, because excretion of colchicine is slow.

NOTES Autumn Crocus is the commercial source of the drug colchicine, which is used as a suppressant for gout and rheumatism, and, in plant genetics, to induce the doubling of chromosomes. It is also being investigated for treating cancerous tumors. It is grown commercially in central and southern Europe and northern Africa. The genus name comes from Colchis on the Black Sea, where the plant is abundant. Autumn Crocus should not be confused with the spring-blooming crocuses of the genus *Crocus*. These belong to the iris family (Iridaceae).

Fig. 137. Belladonna (*Atropa belladonna*). R. & N. TURNER

**Belladonna, or Deadly
Nightshade**
(Nightshade Family)

Atropa belladonna
(Solanaceae)

QUICK CHECK A branching, herbaceous perennial with oval, pointed leaves, tubular purplish flowers, and black, shining berries. Entire plant, particularly berries, contains toxic alkaloids; potentially fatal, especially in children.

DESCRIPTION The plant is a smooth, or slightly hairy, herbaceous perennial with upright, branching stems up to 1.5 m (5 ft) high. The leaves are short-stalked, and usually arranged alternately or in pairs with one leaf much smaller than the other. Generally oval, and pointed, they grow up to 20 cm (8 in.) long. The flowers, borne singly from the leaf axils or forks of branches, are tubular, up to 2.5 cm (1 in.) or more long, widening into 5 lobes at the mouth. They are commonly dull brownish purple to pale blueish purple, violet, or greenish. The fruit, at first green and then red, ripens to a shiny, black or dark purple berry partially enclosed by the 5-lobed calyx.

OCCURRENCE A native of Europe and Asia Minor, Belladonna is sometimes grown as a garden ornamental and herbal plant in North America.

TOXICITY The entire plant contains varying quantities of tropane alkaloids: atropine, hyoscine (also known as scopolamine), and hyoscyamine, as well as a number of other toxic components. Atropine is said to be found throughout the plant, hyoscine in the roots, and hyoscyamine in the flowers, fruits, and seeds. Most cases of human poisoning involve children eating the berries, but adults also occasionally consume concoctions of the leaves or the berries, sometimes mistaking them for edible types such as Bilberries. Symptoms of poisoning include abdominal pain, vomiting, dry mouth, fever, flushed skin, visual disturbances, disorientation, dilated pupils, and rapid pulse. Hallucinations sometimes occur, especially in children. In severe cases, convulsions, coma, and death from respiratory failure may result. Fatalities are rare, but as few as three berries have been reported to be lethal to a child. The plant is seldom eaten by animals, but deaths in livestock are known.

TREATMENT Induce vomiting, or perform gastric lavage; administer activated charcoal and a saline cathartic; sedation may be required, but avoid use of preparations containing morphine or opiates because they reinforce the actions of atropine. In severe cases, a slow intravenous administration of physostigmine is suggested (0.5 mg initially in children; 2 mg for adults), repeated at intervals of 30–60 minutes up to 6 mg for adults and 2 mg for children. See under Jimsonweed in Chapter III (p. 118) for further details.

NOTES The name, Belladonna, comes from the Italian *bella*, beautiful, and *donna*, lady. Women of fashion used to drop juice from the berry into their eyes to dilate their pupils and make them, supposedly, more attractive. The leaves and fleshy root of Belladonna, and the alkaloids they contain, have been valued for centuries as pain killers and sedatives. Today, when used in correct doses, Belladonna alkaloids have many applications; for example, contolling spasms of the urinary tract and excess motor activity of the digestive tract, correction of hypotension associated with slow heart rate, and dilation of the pupil of the eye for opthalmic diagnosis and treatment. The same or related alkaloids are found in other members of the nightshade family, including Henbane (p. 197), Black Nightshade (p. 125), and Jimsonweed (p. 118). All of these should be considered highly toxic.

Cardinal Flowers
(Lobelia Family)

Lobelia cardinalis and *L. fulgens*
(Lobeliaceae)

Fig. 138. Cardinal Flower (*Lobelia fulgens*), cultivar.
R. & N. TURNER

QUICK CHECK Tall herbaceous perennials with showy, bright red flowers; entire plants toxic, potentially fatal.

DESCRIPTION Green or reddish colored, sometimes branching, herbaceous perennials up to 1 m (3 ft) or more tall. If cut or bruised, the plants exude a white sap and give off an unpleasant, acrid odor. The leaves are alternate and lance-shaped, with undulating, slightly toothed margins, and bases running down the stem. The flowers are borne in dense, terminal clusters. They are bright red (occasionally pure white), and irregular, with a tubular portion, a 2-cleft upper lip and 3-cleft lower lip. The fruits are 2-celled capsules containing numerous seeds. *Lobelia fulgens* has deep red foliage.

OCCURRENCE Native to moist areas of central, eastern, and southern North America, Cardinal Flowers are often grown in gardens as ornamentals, for their brilliant red flowers. There are many other wild and cultivated species of *Lobelia*, including short or trailing, blue-flowered species commonly grown as borders and in hanging baskets; all should be considered toxic. Indian Tobacco (*Lobelia inflata*) is discussed separately (p. 114).

TOXICITY The entire plant is poisonous, containing a mixture of pyridine alkaloids, particularly alpha lobeline, which are chemically and functionally similar to Tobacco nicotine (p. 218). Symptoms of poisoning include nausea, vomiting, salivation, abdominal pain, diarrhea, headache, sweating, dilation of the pupils, lack of coordination, confusion, paralysis, lowered temperature, initial high blood pressure followed by low blood pressure, rapid or irregular heart beat, and in severe cases, convulsions, collapse, coma, and death. The sap can irritate the skin. Livestock are also susceptible to poisoning by this plant.

TREATMENT Do not induce vomiting or perform gastric lavage if vomiting has occurred or if seizures are imminent; i.v. diazepam may be used to sedate patients, then give an oral doze of activated charcoal and a saline cathartic (repeat every 6 hours for 24 hours); general supportive treatment for symptoms; artificial respiration and oxygen may be required; atropine may help to combat slow heart beat; treat convulsions with i.v. diazepam or phenobarbital.

NOTES Cardinal Flower (*L. cardinalis*), like its relative Indian Tobacco, has been used in American folk medicine for a variety of ailments, including syphilis and worms. However, since overdoses resulted in severe illness and sometimes death, its use was largely discontinued by the last century.

Celandine
(Poppy Family)

Chelidonium majus
(Papaveraceae)

Fig. 139. Celandine (*Chelidonium majus*). FRANK FISH

QUICK CHECK Yellow-flowered herb, with reddish, acrid juice; entire plant poisonous, occasionally fatal; juice may irritate the skin.

DESCRIPTION An erect, branched biennial or perennial herb, with smooth or slightly hairy stems up to 1.2 m (4 ft) high, with leaves deeply divided into oval segments with irregularly scalloped edges. The flowers, borne throughout the summer in loose, terminal clusters, are bright yellow, 4-petaled, with greenish yellow stamens, and up to 2.5 cm (1 in.) across. There are some double-flowered varieties. The elongated fruiting capsules contain white-tipped black seeds. When cut, the stems exude a yellow-orange latex, or sap, which turns red on exposure to air.

OCCURRENCE A native of Eurasia, Celandine is found as a garden plant in various parts of North America, and also occurs as an escape in some localities, particularly around old gardens.

TOXICITY This plant contains many toxic isoquinoline alkaloids, most notably chelidonine and a-homochelidonine, both of which are chemically related to papaverine, found in Poppy species. Fortunately, the plant is rarely eaten because of its pungent, unpleasant smell and acrid taste. However, children have been poisoned from it, and deaths in both humans and livestock have been reported. Symptoms may not develop for 12 or more hours. They include drowsiness, headache, salivation, and thirst, followed in about six hours by fever, vomiting, diarrhea, and in severe cases, coma and heart failure. The seed capsules are especially potent, and a herd of cattle in Britain were severely poisoned, some fatally, from eating the plants at the fruiting stage. The latex can cause irritation and blistering of the skin.

TREATMENT Induce vomiting, or perform gastric lavage; follow with activated charcoal and a saline cathartic; treat for symptoms; check fluid and electrolyte balance.

NOTES In England, this plant is referred to as Greater Celandine, and an unrelated flower, a Buttercup (*Ranunculus ficaria*), is called Lesser Celandine. Greater Celandine is also known as Rock Poppy, or Wart Wort. This last name alludes to the former use of the latex to elimi- nate warts and corns. It was also used to remove parasites and to treat certain eye disorders, but the treatments often resulted in further irritation and soreness, and have been largely discontinued.

Corn Cockle
(Pink Family)

Agrostemma githago
(Caryophyllaceae)

Fig. 140. Corn Cockle (*Agrostemma githago*).
MARY W. FERGUSON

QUICK CHECK Erect, grayish colored annual with thin, pointed leaves and showy purplish pink flowers; entire plant contains saponins; seeds can contaminate grain; potentially fatal in large quantities.

DESCRIPTION An erect annual, from 30 cm to 1 m (1–3 ft) tall, with a simple or sparsely branched stem and narrow, pointed leaves. The stem and leaves are covered with white hairs, giving the plant a grayish green coloring. The leaves, up to 12 cm (5 in.) long, arise in pairs, clasping the stem at their base. The purplish pink, 5-petaled flowers are borne singly at the ends of the stems, and are about 2.5 cm (1 in.) across, with a bulbous base and narrow, pointed sepals that extend beyond the petals. The fruiting capsules dry to a light brown and open by 5 teeth to reveal many black seeds, each about 3 mm (0.1 in.) long, with a pitted surface.

OCCURRENCE Corn Cockle is a weedy plant native to Europe, and widely established in North America as a weed of cultivated grain fields and waste ground. It is sometimes grown in gardens as an ornamental annual.

TOXICITY The entire plant contains bitter-tasting, soap-like, colloidal glycosides called saponins (mainly githagenin), which do not dissolve in water but are suspended, giving the water a frothing or lathering action. When the seeds contaminate grain, they can cause a chronic form of poisoning, githagism, with symptoms of drowsiness, yawning, weight loss, digestive disturbances, weakness, and even death if the contaminated diet is continued. Acute poisoning is rare, because the plant and its seeds are bitter tasting, and highly contaminated flour has a grayish color and unpleasant odor as well as a bad taste. Acute symptoms of poisoning include severe stomach pain, vomiting, diarrhea, dizziness, weakness, and slow breathing. If githagenin enters the bloodstream, it causes the breakdown of red blood cells. Animals of all types, especially poultry, are susceptible to Corn Cockle poisoning, especially when the seeds contaminate their food.

TREATMENT Induce vomiting, or perform gastric lavage; treat for symptoms. If home-grown or home-ground grains are being used, check for contamination by Corn Cockle seeds.

NOTES All members of the pink family contain saponins, but few except Corn Cockle have been associated with human poisoning. Soapwort, or Bouncing Bet (*Saponaria officinalis*) and Cow Cockle (*S. vaccaria*) are two relatives which are notable for their high saponin content, and should be treated as poisonous, but there are few documented cases of poisoning that implicate them. The main danger from Corn Cockle is in the seeds contaminating grain, especially wheat. Since highly contaminated grain cannot be sold, this weed has caused severe economic losses to North American grain growers. However, with modern agricultural methods and the use of weed killers, Corn Cockle is not as common as formerly, and is less of a problem.

Daffodil and Narcissus
(Amaryllis Family)

Narcissus spp.
(Amaryllidaceae)

Fig. 141. Daffodil (*Narcissus* cultivar). R. & N. TURNER

QUICK CHECK Showy, yellow or whitish flowered spring bulbs; entire plants toxic, but seldom fatal.

DESCRIPTION There are many varieties of Daffodil and Narcissus. They grow from onion-like bulbs, and have long, narrow, often somewhat fleshy basal leaves, and showy flowers born singly or in small clusters at the end of tall, erect, fleshy stalks. They are usually yellow, or whitish, with a characteristic flaring tube emerging from the center of 6 spreading petals. Some varieties are very fragrant.

OCCURRENCE The wild precursors of the many cultivated forms of *Narcissus* are native to Europe and North Africa. These spring-flowering plants are grown in gardens throughout temperate North America, and are also found as house plants and cut flowers.

TOXICITY The entire plants, particularly the bulbs, contain toxic alkaloids (narcissine, or lycorine, and galantamine) and a glycoside (scillaine, or scillitoxin). The bulbs are sometimes mistaken for onions and eaten raw or cooked. They cause dizziness, abdominal pain, nausea, vomiting, and sometimes diarrhea. Trembling, convulsions, and death may occur if large quantities are consumed, but usually recovery occurs within a few hours.

TREATMENT Induce vomiting, or perform gastric lavage; follow with activated charcoal and a saline cathartic; treat for symptoms; replace fluids and check electrolyte balance.

NOTES Flowering bulbs of the same family, *Amaryllis, Hippeastrum,* and *Crinum,* all widely grown in gardens in warm areas and as potted house flowers throughout North America, are reported to contain narcissine and related alkaloids, and should also be considered dangerous. Snowdrop (p. 212) is another relative known to be highly toxic. In the Netherlands, cattle were fatally poisoned after being fed *Narcissus* bulbs during food scarcity in the Second World War.

Fig. 142. Orange-centered Daffodil (*Narcissus* cultivar). R. & N. TURNER

Delphiniums and Larkspurs
(Buttercup Family)

Delphinium spp.
(Ranunculaceae)

Fig. 143. Delphinium (*Delphinium* cultivar).
R. & N. TURNER

QUICK CHECK Herbaceous annuals or perennials with deeply lobed, palmate leaves and (usually) blue, spurred flowers in elongated clusters; entire plants highly toxic, potentially fatal.

DESCRIPTION There is considerable variation in the size and form of *Delphinium* species. A common distinction is that the tall, leafy, clumped forms are called Delphiniums, whereas the short single-stemmed, sometimes annual types are called Larkspurs. The leaves of both types are usually deeply palmately lobed and long-stalked. The flowers are showy, borne in dense, elongated clusters, or racemes. The most common flower color is blue, ranging in shade from light sky blue to deep purplish blue; there are also white, mauve, and pink shades. The flowers have flaring petals, and each has a characteristic spur projecting backward from the upper part. The fruits are a cluster of dry, many-seeded follicles, each splitting along the side.

OCCURRENCE Many hybrid perennial Larkspurs, or Delphiniums are grown in gardens throughout temperate North America. In addition, there are many native species growing in the wild in different parts of the continent.

TOXICITY Delphiniums and Larkspurs contain a group of toxic alkaloids (including delphinine, delphineidine, and ajacine) similar to the aconitines found in Monkshood (p. 204). The young plants and the seeds contain the highest concentrations. The plants can cause severe and often fatal poisoning of animals, especially range animals in the wild, but since humans seldom consume them, there are few reports of people being poisoned. In one instance, a man ingested Delphinium leaves and seeds and five hours later developed symptoms of nausea, vomiting, abdominal pain, blurred vision, dry skin and mouth. Restlessness, agitation, and dilation of the pupils lasted for 12 hours, after which recovery occurred.

TREATMENT Induce vomiting, or perform gastric lavage; follow with activated charcoal and a saline cathartic; maintain fluid and electrolyte balance; monitor heart rate, blood pressure, and breathing; administer atropine for low heart rate, and if necessary, dopamine for low blood pressure; control convulsions with i.v. diazepam; oxygen and artificial respiration may be required.

NOTES There are several other related plants in the Buttercup Family that are toxic. As well as Baneberry (p. 99), Buttercups (p. 103), Clematis (p. 173), Hellebore (p. 195), and Monkshood (p. 204), there are others found in the wild and grown as garden flowers: Anemones, Marsh Marigolds, and Columbines. Most contain irritant protoanemones (see discussion, p. 104), and Columbine contains alkaloids similar to those of Delphinium and Monkshood. Another poisonous garden plant in this family, with showy yellow or crimson flowers, is Pheasant's Eye (*Adonis* spp.). The whole plant contains digitalis-like glycosides, and possibly a protoanemonin-like irritant. Poisoning by this species has not been reported in North America.

Four-O'Clock, or Marvel of Peru
(Four-O'Clock Family)

Mirabilis jalapa
(Nyctaginaceae)

Fig. 144. Four-O'Clock (*Mirabilis jalapa*). R. & N. TURNER

QUICK CHECK Herbaceous perennial with showy, tubular flowers; roots and seeds poisonous, but apparently not fatal.

DESCRIPTION Erect, branching, herbaceous perennial up to 1 m (3 ft) tall, with dark green, opposite, stalked leaves that are oval, up to 15 cm (6 in.) long, and pointed at the tip. The flowers are white to red, yellow, or striped, with a long tubular portion, 5-lobed and flaring at the top, and about 2.5 cm (1 in.) across. The name derives from their usual flowering time: late afternoon, and right through the night. The fruit is a leathery, 5-ribbed, 1-seeded achene.

OCCURRENCE A native of tropical Central and South America, Four-O'Clock is commonly grown as an ornamental in North American gardens. In parts of the southern United States, it has escaped cultivation.

TOXICITY The roots and seeds have caused acute stomach pain, vomiting, and diarrhea in children who have eaten them. The toxic agent is apparently unknown.

TREATMENT Induce vomiting, or perform gastric lavage; treat for symptoms.

NOTES According to Lewis and Elvin-Lewis (1977), Four-O'Clock is being investigated in cancer therapy. It has also been used as a purgative.

Foxglove
(Figwort Family)

Digitalis purpurea
(Scrophulariaceae)

Fig. 145. Foxglove (*Digitalis purpurea*). R. & N. TURNER

QUICK CHECK Biennial with large, oval basal leaves and tall flowering stalks with many purplish (or white) tubular flowers, maturing from bottom to top of the stalk; entire plant contains potent glycosides affecting the heart; potentially fatal.

DESCRIPTION An erect biennial (or occasionally perennial) herb, with flowering stem up to 1.5 m (5 ft) high. In the first year the leaves form a rosette at ground level. The basal leaves are large, oval, pointed at the tip, prominently veined, and finely toothed around the margins. Especially on the undersides, they are covered with short, soft hairs, giving the plant a grayish green cast. The flowering stems bear leafy bracts similar to, but smaller than, the basal leaves. The flowers are borne in a terminal, many flowered spike, itself up to 60 cm (2 ft) long, with the tubular blossoms crowded along one side, hanging or drooping, and maturing in sequence from bottom to top. They are purplish pink, or sometimes white or yellowish, with a lower, paler colored, projecting lip which is usually spotted with purple or brown.

The fruits are oval capsules which dry to a light brown and split open to reveal numerous tiny, dark brown seeds.

OCCURRENCE Foxglove is a hardy plant, originating in southern Europe and central Asia and now widely distributed in North America. It is grown in gardens and is found as an escape along roadsides, in fields, and in burned or logged over areas. It is also cultivated in quantity as a medicinal drug plant.

TOXICITY Foxglove contains potent cardiac or steroid glycosides, including digitoxin, digoxin, digitalin and related compounds. These compounds break down in the body to a sugar and a non-sugar component (aglycone). Acting together, they have a direct influence on the muscles of the heart. Even when dried or boiled, Foxglove retains its active compounds. Children have been poisoned from eating the flowers, sucking the nectar, or even drinking water from vases containing the flowers. Occasional poisonings in adults occur from misuse of herbal preparations of the plant, misidentification of the plant, or overdoses of the heart drug digitalis.

Symptoms of poisoning include pain in the mouth and throat, nausea, vomiting, abdominal cramps, diarrhea, severe headache, irregular heartbeat and pulse, tremors, and in severe cases, convulsions and death due to cardiac arrest. Animals seldom eat Foxglove, but it may be eaten in times of food scarcity or as a hay contaminant, sometimes with fatal results.

TREATMENT Induce vomiting, or perform gastric lavage; follow with activated charcoal and saline cathartic; transvenous cardiac pacing may be required; monitor serum electrolytes and heartbeat. Administer atropine to restore normal sinus rhythm and increase heart rate; phenytoin or lidocaine is recommended for rhythm

Fig. 146. Foxglove, basal leaves. R & N. TURNER

Fig. 147. Foxglove, flowers. R & N. TURNER

disturbances; for severe hyperkalemia, administer glucose and insulin, or sodium polystyrene sulfonate (Canadian Pharmaceutical Association 1984). Lampe and McCann (1985) note that Fab fragments, digitalis-specific antibodies, though not yet widely available, may become the chosen treatment for Foxglove or digitalis poisoning.

NOTES Foxglove leaves closely resemble those of Comfrey (*Symphytum officinale*), especially when young, and confusion between these two plants has resulted in poisoning since some people make herbal tea from Comfrey. This practice is not recommended, due to the potential presence of liver-damaging pyrrolizidine alkaloids in some forms of Comfrey, but those who do drink Comfrey tea should learn to distinguish between these two common garden plants. Foxglove leaves are finely toothed along the edges, whereas Comfrey leaves, though hairy, are smooth-edged.

Digitalis, the dried, powdered leaf of Foxglove, is a highly important drug in modern heart treatment, but the doses must be carefully monitored, and people taking this medication should take care to keep it out of reach of children.

Glory Lily, or
African Climbing Lily
(Lily Family)

Gloriosa superba
(Liliaceae)

Fig. 148. Glory Lily (*Gloriosa superba*). R. & N. TURNER

QUICK CHECK Slender herbaceous plant or vine from a tuberous rootstock, with showy, yellow or red lily-like flowers with swept-back petals; all parts extremely poisonous, potentially fatal.

DESCRIPTION Slender, herbaceous, perennial vine up to 1.5 m (5 ft) long, growing from a thick, tuberous rootstock. The leaves are alternate or almost opposite, lanceolate or elliptical, and up to 15 cm (6 in.) or more long, with a tendril-like tip that allows the plant to climb. The large, striking flowers are yellow at first, changing to orange and then to bright crimson. They are borne on long stalks, and each has six "petals" that are crinkled along the edges and spreading, or reflexed, with the stamens and single green pistil projecting down. The fruits are large, oblong capsules containing bright red seeds.

OCCURRENCE Native to tropical Africa and Asia, Glory Lily is frequently grown in gardens of the southern United States, and as a potted plant, mainly in greenhouses, elsewhere in North America.

TOXICITY The entire plant is toxic, particularly the tubers. It contains the alkaloid colchicine and other compounds related to those found in Autumn Crocus. Symptoms of poisoning include intense burning and numbness of the mouth and throat, thirst, nausea, and vomiting. Abdominal pain and severe diarrhea may follow after a delay of two or more hours. Bloody urine, reduced urine output, difficulty in breathing, convulsions, and sometimes death may also result.

TREATMENT Induce vomiting or perform gastric lavage, followed with activated charcoal and a saline cathartic; maintain fluid and electrolyte balance; monitor blood pressure, kidney function, serum potassium, respiration, and heart rate; analgesics and atropine may be required, and artificial respiration may be necessary. Symptoms can last a long time, because excretion of colchicine is slow.

NOTES A related species, *Gloriosa rothschildiana,* is also occasionally grown in southern gardens. Because of its colchicine content, Glory Lily is sometimes used in cancer chemotherapy.

Hellebore, or Christmas Rose
(Buttercup Family)

Helleborus niger and related
   species
(Ranunculaceae)

Fig. 149. Hellebore (*Helleborus niger*). R. & N. TURNER

QUICK CHECK Winter and early spring blooming herbaceous perennial with deeply divided leaves and large whitish or purplish flowers. Entire plant contains Foxglove-like cardiac glycosides; poisoning rare but may be fatal.

DESCRIPTION Herbaceous perennial with thick, fibrous roots and long-stalked, compound leaves, the segments coarsely toothed. The flowering stems, up to 45 cm (18 in.) high, are simple or forked, bearing small leaves or bracts. The flowers bloom in winter and early spring. They are large and nodding, with five white, pinkish, or dark reddish purple, petal-like sepals. The fruit is a small capsule containing many glossy, black seeds.

OCCURRENCE Hellebore is native to Europe, and is widely cultivated. It and its hybrids are grown in gardens throughout temperate North America, and are valued for their winter and early spring flowering times. In some places it has escaped from cultivation.

TOXICITY Plants in the genus *Helleborus* contain toxic glycosides (helleborin, helleborein, and helleborigenin), closely related to those of Foxglove (p. 192). In the past, human and animal poisoning frequently occurred from misuse of herbal preparations of Hellebore as purgatives, local anaesthetics, or abortives, or to rid the skin of parasites. Because of the health risks its medicinal use has been largely abandoned, and instances of poisoning by it are rare. Symptoms, which may be delayed depending on the quantity ingested, include pain of the mouth and abdomen, nausea, vomiting, cramps, diarrhea, visual disturbance, loss of appetite, slow heart beat, and possible abortion in pregnant women. Drinking milk from poisoned cows can cause vomiting and diarrhea.

TREATMENT Induce vomiting, or perform gastric lavage; follow with activated charcoal and a saline cathartic; monitor heartbeat and serum potassium; atropine or transvenous pacing may be required to correct heart rate.

NOTES Hellebore and its relatives are sometimes confused with another plant, known variously as False Hellebore, Indian Hellebore, or Green Hellebore (*Veratrum viride*). This plant in the lily family is unrelated to the Hellebores, but is also very poisonous (see p. 110).

Henbane
(Nightshade Family)

Hyoscyamus niger
(Solanaceae)

Fig. 150. Henbane (*Hyocyamus niger*). WALTER H. LEWIS

QUICK CHECK Hairy, herbaceous weed with disagreeable odor, large basal leaves, and funnel-shaped yellowish or mauve flowers borne along upper stems; entire plant toxic, but seldom fatal in humans; dangerous as a medicinal herb.

DESCRIPTION Hairy, somewhat sticky annual or biennial weed about 60 cm (2 ft) high, with a woody stem base. The entire plant has a strong, disagreeable smell. The basal leaves are stalked, up to 20 cm (8 in.) long, and with large, irregular teeth on the margins. The stem leaves are stalkless, alternate, and unevenly lobed. The flowers, produced singly from the stems just above the leaves, are subtended by leafy bracts. The flowers are funnel-shaped, about 2.5 cm (1 in.) across, and greenish yellow to whitish, or occasionally mauve, conspicuously marked with a network of dark purplish veins. In fruiting, the calyx grows around the fruit and hardens into five pointed spines. The exposed cap of the ripened fruiting pod comes off, revealing numerous seeds.

OCCURRENCE A native of Europe, Henbane is a widespread weed of waste places in the northeastern United States, and is found sporadically across the continent, sometimes being grown for herbal use.

TOXICITY All parts of the plant, particularly the roots and seeds, are poisonous. They contain tropane alkaloids of the same group as those in Jimsonweed (*Datura* spp.) and Belladonna (*Atropa belladonna*), namely hyoscyamine, hyoscine (also known as scopolamine), and atropine. Symptoms of poisoning, which may persist for 12 hours or more, include dry mouth, dry skin, fever accompanied by rash, blurred vision, dilated pupils, rapid heartbeat, excitement, dizziness, delirium, confusion, headache, and, especially in children, hallucinations. Most incidents of accidental poisoning in humans are from earlier times in England and Europe, when Henbane was commonly used medicinally. Occasional cases are still reported, either from misidentification of the plant or its roots as edible types, or from its misapplication as a medicine. The plant is unattractive to animals, but sometimes cattle and other livestock are poisoned by it, with symptoms similar to those of human poisoning.

TREATMENT Induce vomiting, or perform gastric lavage; follow with activated charcoal and a saline cathartic. For further details of treatment, see under Belladonna (p. 183).

NOTES Henbane has long been known in Europe as a medicinal plant. It has been used for eye disorders, rheumatism, and as a sedative. In the Middle Ages it was also used in sorcery.

Iris, or Flag
(Iris Family)

Iris pseudacorus, I. × germanica,
and related species
(Iridaceae)

Fig. 151. Iris (*Iris* cultivars). R. & N. TURNER

QUICK CHECK Erect, perennial herbs growing from rhizomes (or bulbs), with showy, colorful flowers in a distinctive formation: three outer spreading "petals" and three erect inner ones. Entire plants, especially the rhizomes, are poisonous, but seldom fatal in humans.

DESCRIPTION Irises, or Flags, are perennial herbs growing from thick rhizomes or bulbs, often occurring in dense clumps or patches. The stems are erect, either simple or branched, and the leaves are mostly basal. They are smooth-edged, long, narrow, and pointed, grass-like in some, broader in others. They often form two opposite ranks, sheathing the stem at the base, and becoming flattened vertically in one plane in a fan-like formation. The showy, short-lived flowers are borne singly or in small clusters at the ends of the stems. The formation of the flowers is distinctive, with three outer "petals" which are hanging or reflexed and three inner ones usually erect and often arched. The flowers of *Iris pseudacorus*, Yellow Flag, are bright yellow; those of *I. × germanica* (a hybrid), the common garden Iris, come in a wide variety of colors, including white, yellow, blue, purple, reddish, brownish, and variegated. Other species have yellow, blue, or purple flowers. The fruits are oblong, many-seeded capsules.

OCCURRENCE There are some 200 species of *Iris*, most native to the North Temperate Zone. Several species grow wild in parts of North America, and many types are widely grown as garden ornamentals. *Iris pseudacorus* is a widespread garden escape, found commonly in ditches and marshes and along lake edges.

TOXICITY All parts of wild and cultivated *Iris*, especially the rhizomes, are poisonous. They contain as yet unknown toxic compounds, possibly an irritant resin and a glycoside (called iridin, irisin, or irisine). If eaten, Iris can cause severe digestive upset, with abdominal pain, nausea, vomiting, diarrhea, and fever. Humans are seldom poisoned, but there are numerous cases of fatalities in livestock from Iris poisoning, and children should be warned against chewing on the leaves or fleshy portions of wild or cultivated Iris plants. Gardeners should know that Iris sap can irritate the skin, sometimes causing blistering.

TREATMENT Induce vomiting, or perform gastric lavage; replace fluids to prevent dehydration; treat for symptoms.

NOTES A major danger from marsh-loving Flags is in the similarity of their leaves to those of Cattail (*Typha latifolia*). The latter plant is often found growing in marshy areas side by side with *Iris pseudoacorus.* Cattail has edible rootstocks, shoots, and leaf bases, but those wishing to use it should be careful not to take Flag instead. Iris has been used medicinally as a purgative.

Lily-of-the-Valley
(Lily Family)

Convallaria majalis
(Liliaceae)

Fig. 152. Lily-of-the-Valley (*Convallaria majalis*).
R. & N. TURNER

QUICK CHECK Herbaceous perennial with smooth-edged, elliptical leaves, prized in gardens for its fragrant, dainty, bell-like, white flowers arranged along one side of the stalk. Entire plant toxic, with digitalis-like symptoms; sometimes fatal.

DESCRIPTION Low, herbaceous perennial growing from creeping underground rootstocks, often forming extensive patches. The stems are upright, bearing 2 oval to elliptical, pointed, bright green leaves, smooth-edged, up to 20 cm (8 in.) long, and tapering to a clasping stalk. The white, delicate flowers are fragrant, bell-like, and globular. They are borne in a cluster of 6–12 along one side of a slender stalk, usually no more than 10 cm (4 in.) long. The fruits are fleshy, orange-red berries.

Fig. 153. Lily-of-the-Valley, fruits. R. & N. TURNER

OCCURRENCE A native of Eurasia, Lily-of-the-Valley is frequently cultivated in gardens throughout temperate North America. It is also found as a garden escape in eastern North America. A related species, *Convallaria montana,* is native to the montane woods of North Carolina, Tennessee, Virginia, and West Virginia.

TOXICITY The entire plant is poisonous, containing a large group of digitalis-like cardiac glycosides, including convallotoxin, convallarin, and convallamarin. Convallotoxin is one of the most toxic of all naturally occurring substances that affect the heart. Some of the glycosides are saponins which, together with volatile oils present in the plant, act as digestive tract irritants. Symptoms of poisoning are burning pain in the mouth and throat, heavy salivation, nausea, vomiting, abdominal pain, cramping, diarrhea, headache, dilated pupils, cold clammy skin, and slow irregular heartbeat, sometimes leading to coma and death from heart failure. Depending on the quantity consumed, there may be a latent period before some of these symptoms are experienced. Children have been poisoned, some fatally, from eating the berries or chewing on the leaves. In one instance, the leaves were mistaken by adults for wild garlic leaves and were cooked in soup; those eating it became hot and flushed, and developed headache and hallucinations. Even water in which the cut flowers have been kept can be dangerous to drink.

TREATMENT Induce vomiting, or perform gastric lavage; follow with activated charcoal and saline cathartic; transvenous cardiac pacing may be required; monitor serum electrolytes and heartbeat. Administer atropine to restore normal sinus rhythm and increase heart rate; phenytoin or lidocaine is recommended for rhythm disturbances; for severe hyperkalemia administer glucose and insulin, or sodium polystyrene sulfonate.

NOTES Lily-of-the-Valley was formerly employed medicinally as a heart stimulant and diuretic, but today has been largely superseded by digitalis from Foxglove, which is more reliable.

Fig. 154. Lupine (*Lupinus* cultivar). R & N. TURNER

Lupines
(Bean, or Legume Family)

Lupinus spp.
(Fabaceae, or Leguminosae)

QUICK CHECK Perennial or annual herbs or woody-based shrubs with palmately compound leaves, and smallish, pea-like flowers in elongated, terminal clusters, and flat, often hairy or silky, pea-like fruiting pods.

DESCRIPTION Most Lupines are herbaceous perennials, but some are annuals and some are shrubby, with a woody base. The leaves are alternate, and palmately compound, with 5–15 elongated, pointed leaflets radiating from the end of the leaf stalk. The flowers are pea-like, crowded in showy, elongated terminal clusters, with the most mature flowers at the bottom and the last-opening at the top. Attractively colored, they range from blue to yellow, pink, white, or variegated. The pea-like pods are flattened, brownish or blackish when ripe, and often hairy or silky. They contain several flattened seeds.

OCCURRENCE There are about 200 species of Lupine, most native to western North America and the Mediterranean region. Many varieties are grown as garden ornamentals, the most common being forms of *Lupinus polyphyllus*. These often become naturalized around old gardens and homesteads, and some lupines are now being sown along highways for roadside stabilization.

TOXICITY Lupines contain a large number of potent alkaloids, including lupinine, lupanine, anagyrine, and sparteine, which act on the nervous system. Some lupines also contain enzyme inhibitors and other toxins. The toxicity of the plants varies with species, time of year, and part of the plant. The seeds and pods, and the young leaves and stems in spring are the most dangerous. Many species are known to be toxic to grazing livestock, causing deaths and birth deformities, and some have been harmful to humans as well. The main danger of human poisoning is in eating the green, unripe seeds or pods, which may be mistaken for peas or beans, or the ripe seeds, whose pods may attract children because of their silky covering. The symptoms of poisoning are similar to those of Laburnum (p. 158), but not as severe. They include vomiting, excessive salivation, nausea, dizziness, headache, and abdominal pain. Slowed breathing and heartbeat may also occur. In extreme cases, death through respiratory failure may result, but this is rare.

TREATMENT Most cases of lupine poisoning are mild, and treatment is rarely required. If symptoms are severe, follow treatment for Laburnum (*Laburnum anagyroides*) (p. 158).

NOTES One species of lupine (*Lupinus latifolius*), which formed the main forage in an area heavily browsed by milk goats in Trinity County, California, was linked in 1981 to deformities in a human baby and in puppies when the mothers drank the goats' milk during their pregnancies, and to deformities and stillbirths in the goats during the same time period. The seeds and dried leaves of this species are known to have a high content of the quinolizidine alkaloid anagyrine.

Marijuana, or Hemp
(Hemp Family)

Cannabis sativa
(Cannabaceae)

Fig. 155. Marijuana (*Cannabis sativa*). MARY W. FERGUSON

QUICK CHECK Weedy, erect annual with palmately compound leaves with narrow, toothed leaflets. Entire plant contains toxic resins, which produce mind-altering symptoms ranging from mild euphoria to confusion, depression, panic, memory loss, and coma, depending on the amount and nature of the product taken, the plant strain, and the person. Injection of hash oil can be fatal.

DESCRIPTION Erect, herbaceous annual, with fluted or angled stems often 1.2–5 m (4–16 ft) tall. The leaves on the lower part of the plant are borne from thickened nodes at intervals along the stem. They are palmately compound, with usually 7 narrow, pointed, coarsely toothed leaflets, from 5–15 cm (2–6 in.) long, arising from a single point at the end of the stalk like fingers on a hand. Male and female flowers are produced on different plants. They are inconspicuous, hidden among small leaves at the ends of branches. After flowering, the male plants turn yellow and die; the female plants remain dark green for about a month longer and produce small fruits ("hemp seeds").

OCCURRENCE It is difficult to know where to place Marijuana in this book, because it is grown (illegally) both indoors and outdoors for its "narcotic" effect, and is also a persistent weed of roadsides and waste places in parts of central and eastern North America. In some parts of the world a tall strain of this plant is cultivated for fiber production. It is also grown for its oil-rich seeds, as well as for production of the narcotic cannabis.

TOXICITY Marijuana plants, both fresh and dried, contain toxic resins including tetrahydrocannabinol (THC). The concentration of these compounds varies considerably with the strain of the plant, part of the plant, its growth stage, how it is consumed, and the individual taking it. This variability comes mainly from the instability of some of the constituents. Smoking Marijuana cigarettes, or "joints," is about three times more powerful than ingestion. In small amounts THC and its related compounds produce mild visual effects, brief mood elevation, and sedation; concentrated doses can cause confusion, depression, paranoia, panic, memory loss, rapid heart beat, high blood pressure, and coma, depending on the amount and nature of the product taken, the plant strain, and the person. Smoking marijuana can cause throat irritation and coughing, as well as dry mouth, eye irritation, nausea, and vomiting. Injection of hash oil (a crude extract of hashish, or marijuana resins) can be fatal. The plant is bitter tasting and is seldom touched by animals, but it has caused fatal poisoning of horses and mules in Greece. Dogs fed Marijuana products by their owners have shown many serious symptoms—in some cases muscular tremors, convulsions, prostration, and coma—but recovered within 24 hours in the cases reported.

TREATMENT Induce vomiting, or perform gastric lavage; follow with activated charcoal and saline cathartic. Maintain a quiet, subdued environment for victim; sedation in the form of oral or i.v. diazepam may be required. For poisoning from intravenous injection of cannabis, treat for hypotension; fluid replacement and vasopressors may be required; ensure adequate urine output (mannitol and furosemide are suggested).

NOTES Cannabis is one of the earliest known narcotics. It was formerly used in human and veterinary medicine as a sedative and hypnotic, but now is used mainly for psychedelic effects. It is taken in the form of a "tea," by chewing the plant parts, or by smoking the dried leaves or extracted resins. Those wishing to indulge in Marijuana smoking should be advised that the effects can be highly variable and unpredictable, and are not necessarily pleasant.

Monkshood, Aconite, or Wolfsbane
(Buttercup Family)

Aconitum napellus and related species
(Ranunculaceae)

Fig. 156. Monkshood (*Aconitum napellus*), foliage.
R. & N. TURNER

QUICK CHECK Herbaceous perennial similar to Delphiniums, with deeply divided leaves and bluish, helmet-shaped flowers in elongated, terminal clusters; entire plant highly toxic and potentially fatal.

DESCRIPTION Perennial herb growing from a dark, tuberous tap root, with usually unbranched stem up to 1 m (3 ft) or more high. The leaves are more or less triangular in outline, but deeply divided into narrow, pointed segments. The flowers are borne in an elongated cluster at the end of the stem, and are blue, blueish mauve, or purple and helmet-shaped, each with an enlarged, hood-like "petal" forming a cap over the rest of the flower. The fruits usually form a tight group of 3 follicles, each splitting along the inside and containing many sharply angled seeds. There are some 100 species of Monkshood native to the North Temperate zone. Flower colors range from blue to yellow to white. These plants differ widely in chemical constituents, but all should be considered poisonous.

OCCURRENCE There are several species of *Aconitum* native to in parts of North America, mostly found in moist, montane areas. *Aconitum napellus* is native to Europe, and is grown as an ornamental in North American gardens, as are several other species.

TOXICITY Monkshood contains several potent closely related alkaloids, especially aconitine and aconine. It is regarded as the most poisonous plant in Great Britian, according to Cooper and Johnson (1984). Its similarity to some species of Larkspur, or Delphinium (*Delphinium* spp.), also toxic and more common than Monkshood, has occasionally resulted in confusion in identification of poisoning causes. People are sometimes poisoned from misuse of herbal preparations containing Monkshood, or from mistaking it for another plant. In one instance, fatal human poisoning occurred when the roots were mistaken for those of Horseradish. The roots have also reportedly been mistaken for celery, and the leaves for parsley. Symptoms include tingling and burning of the lips, tongue, mouth and throat, abdominal

Fig. 157. Monkshood (*A. columbianum*), closeup of flower. IAN G. FORBES

pain, excessive salivation together with intense thirst, severe vomiting, diarrhea, headache, cold feeling, slow heart rate, paralysis, confusion, restlessness, visual disturbances, convulsions, and delirium. Coma and death may occur rapidly from asphyxiation and circulatory failure, but in most cases the patient recovers fully within 24 hours. Animals seldom eat Monkshood, but it does cause poisoning and occasional fatalities in livestock. A fatal dose for a dog has been estimated at 5 g (0.2 oz), for a horse, 350 g (12 oz).

TREATMENT Induce vomiting, or perform gastric lavage; follow with activated charcoal and a saline cathartic; monitor breathing, heart rate, and blood pressure; oxygen may be needed to assist breathing; keep patient warm; maintain fluid and electrolyte balance; heart-stimulating drugs such as atropine may be required; dopamine may be used to treat persistent low blood pressure; i.v. diazepam has been recommended to control convulsions or restlessness.

NOTES The poisonous nature of Monkshood has been known since ancient times in China, India, and Europe and, probably because of its high potency, it has a long history of use as a folk medicine. *Aconitum napellus* is the source of the pharmaceutical drug aconite, which is used mainly as a local analgesic in liniments.

Poppy, Opium, and Its Relatives
(Poppy Family)

Papaver somniferum and related species
(Papaveraceae)

Fig. 158. Opium Poppy (*Papaver somniferum*), showing scored seed capsules. R. & N. TURNER

QUICK CHECK Tall, large-flowered annual, with whitish, mauve, or reddish flowers and globular seed capsules; the numerous, small seeds are harmless in small amounts and used as a condiment; eating other parts of the plant, especially the unripe seed capsules, may cause deep sleep and coma; seldom fatal.

DESCRIPTION Erect, robust annual, up to about 1.2 m (4 ft) tall, with a uniformly blueish green cast. The leaves, up to 25 cm (10 in.) long, are alternate, lobed or toothed, with undulating margins. They are elongated or heart-shaped, with a clasping base. The flowers are borne singly, or two or three together, at the ends of the stems. The flower buds are nodding, the open flowers upright. They are large and showy, ranging in color from white to pale lilac to reddish or orange, often with dark blotches at the base of the petals. The fruit is a large, smooth, spherical or oval capsule, with a scalloped "lid." The small black or white seeds are numerous. The entire plant, especially the unripe seed capsules, contains a milky sap, which exudes when the stem or capsules are cut. There are many variable forms and varieties of this plant, including popular double-flowered ones. Other species of Poppy are also commonly grown as garden flowers, and a few are native to parts of North America. Cultivated species include: Iceland Poppy (*P. nudicaule*), Oriental Poppy (*P. orientale*), Corn Poppy (*P. rhoeas*), and California Poppy (*Eschscholtzia californica*).

OCCURRENCE Although growth of Opium Poppy is restricted by law because of its narcotic properties, it is nevertheless often found in North American gardens (especially old-fashioned gardens), and as an escape in waste places. It originated in Eurasia, and in some areas of the world it is cultivated as a medicinal plant and for its edible seeds. It is also grown for the illicit drug trade, mainly in the Middle East and Asia.

TOXICITY The entire plant contains the drug opium, a crude resin which consists of a mixture of many potent alkaloids, including the well known compounds, morphine and codeine, as well as papaverine and a variety of related substances. The unripe fruiting capsules are the main commercial source of opium, which exudes when the outside of the growing capsule is scored or cut. The ripe seeds contain only minute concentrations of opium; they are considered harmless, and are used as a condiment and for decorating bread. Human poisoning occurs from eating the unripe capsules, and from overdoses of opium or its derivatives, morphine, heroin, and codeine. These drugs are highly addictive, and illegal trafficking of them is a major problem in North America. Opium derivatives mimic naturally occurring compounds—endorphins and encephalins—which are found in the brain and other parts of the neuroendocrine system, and serve as natural pain suppressants, sedatives, and mood elevators under certain conditions.

Symptoms of Poppy poisoning include include initial restlessness, excessive salivation, loss of appetite, stupor, shallow and slow breathing, deep sleep, and coma. Although poisoning is seldom fatal, recovery can be slow. Furthermore, opiates are strongly addictive, and withdrawal from them once addiction has set in is a ghastly experience. Large doses of opium or its derivatives can cause death through respiratory failure. All Poppy species should be considered potentially toxic; they have caused poisoning in livestock, with severe

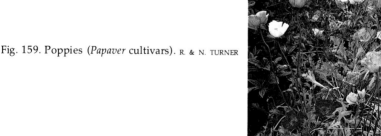

Fig. 159. Poppies (*Papaver* cultivars). R. & N. TURNER

economic losses through slow recovery, and permanently depressed milk yield in lactating females.

TREATMENT Induce vomiting, or perform gastric lavage; follow with activated charcoal and a saline cathartic; monitor breathing, and be prepared to treat depressed respiration if required; narcotic antagonists; nalaxone has been a recommended antidote (1.2 mg for adults, 0.03 mg/kg body weight for children; repeat once if required); strong coffee has also been suggested.

NOTES Opium is possibly the oldest narcotic known; records of its use go back as early as 4000 B.C. It was used as a pain killer and sleep-inducing drug by the ancient Greeks and Romans. By the nineteenth century innumerable patent medicines containing opium became popular in Europe and North America; opium-based teething syrups, cough medicines, sedatives, pain killers, and cures for diarrhea accounted for a major portion of prescription medicines and many people became addicted to these medications. In China, opium smoking became a habit of millions, due to aggressive importation of the drug by British and American merchants. In 1838 a Chinese offical attempted to enforce a ban on the opium trade and precipitated the famous Opium Wars, in which the British retained the right to import opium to China.

Several other members of the poppy family are known to be toxic, including Celandine (*Chelidonium majus*, p. 186), Bloodroot (*Sanguinaria canadensis*, p. 102) and Prickly Poppy (*Argemone mexicana*), a yellowish-flowered, prickly weed commonly occurring in the southern United States. All of these species exude a yellowish to reddish sap when cut. Prickly Poppy seeds are highly toxic and may cause severe poisoning when they contaminate home-ground grain.

Potato (Greens and Sprouts)
(Nightshade Family)

Solanum tuberosum
(Solanaceae)

Fig. 160. Green potatoes (*Solanum tuberosum*).
R. & N. TURNER

QUICK CHECK Potato tubers are a major food, but the sprouts, green tubers, and berries can be dangerous; not usually fatal in humans.
DESCRIPTION Tuber-bearing herbaceous perennials with dark green, com-

pound leaves, and clustered whitish to purplish flowers with reflexed petals and protruding, yellow stamens. The tubers are variously shaped, with different colors and textures depending on their age and variety. Most commonly, they are brownish and ovoid. Red, yellow, and purplish tubers are also grown. The seldom produced fruits are globular, yellow berries.

OCCURRENCE Potatoes are widely grown in North American gardens, and a major produce item in grocery stores everywhere. They are native to South America, but are now a major crop plant in North America, Europe, and the U.S.S.R.

TOXICITY Potato tubers are a major source of food for North Americans. However, the plant is in the same genus as Climbing Nightshade (*Solanum dulcamara*—p. 94) and Black Nightshade (*Solanum nigrum* and *S. americanum*—p. 125), and, like them, contains the glycoalkaloid solanine and related alkaloidal compounds, including some similar to atropine. Most cases of potato poisoning result from eating potato tubers which have turned green through exposure to light, or which have sprouted. In these tubers, the solanine content is many times higher than in normal, properly stored potatoes. The highest concentration of the toxins is in the skin, eyes, and sprouts of the tubers. Infection with potato blight fungus increases the concentration of toxins. Boiling in water reduces, but does not eliminate, the alkaloids from green or sprouted potatoes. (If potatoes have "gone green," they may be stored in the dark for 2–3 weeks until all traces of green have disappeared, then safely eaten.) There is also danger from the berries, which are 10–20 times as high in toxic alkaloids as the normal tubers.

Children are more susceptible to solanine poisoning than adults. Symptoms may be delayed for several hours after ingesting toxic parts. They include: severe abdominal pain, diarrhea, fever, sweating, headache, restlessness, delirium, visual disturbances, convulsions, drowsiness, and coma. There may also be urinary retention and kidney failure, but few fatalities in humans have been reported. Livestock have frequently been poisoned, often fatally, from feeding on green, decayed, or sprouting potatoes, peelings, or fruiting plants.

TREATMENT Induce vomiting, or perform gastric lavage; follow with activated charcoal and a saline cathartic; antacids may relieve digestive distress; maintain fluid and electrolyte balance; monitor blood pressure, heartbeat, temperature, and kidney function; supportive treatment for symptoms, including tranquilizers or anticonvulsants if required. I.v. diazepam is recommended to control excitation or convulsions, and, if necessary, a slow i.v. injection of physostigmine salicylate may be used to reverse convulsions, heart irregularities, and hallucinations. Sponge with tepid water to reduce fever. Recovery may take a week or more.

NOTES One of the most dramatic cases of potato poisoning took place in the late 1970s, when 78 schoolboys were poisoned after eating potatoes from a sack which had been stored improperly for several weeks. After a period of 8–10 hours they experienced abdominal pain, followed by diarrhea. Some also suffered restlessness, delirium, visual disturbances, drowsiness, or coma. Three of the boys became seriously ill, but even they recovered completely within about a month.

Symptoms of solanine poisoning are usually delayed, because the glycoalkaloids are not readily absorbed; first they break down to release free alkamines, which are then absorbed, producing symptoms of drowsiness and dulling of the senses.

Rhubarb (Leaves)
(Buckwheat Family)

Rheum rhabarbarum (formerly
R. rhaponticum)
(Polygonaceae)

Fig. 161. Rhubarb (*Rheum rhabarbarum*). R. & N. TURNER

QUICK CHECK Common perennial garden plant with very large, oval or heart-shaped basal leaves, forming dense clumps. The fleshy, reddish leaf stalks are sour tasting but edible in moderation; the green leaf blades are highly toxic and potentially fatal if eaten.

DESCRIPTION Perennial plant from large, fleshy rootstocks. The leaves are mostly basal and very large, with thick, fleshy stalks that may grow 60 cm (2 ft) or more long (the edible vegetable part of the plant), and green, oval or heart-shaped blades up to 45 cm (18 in.) or more long. The small, numerous, greenish-white flowers are produced in dense clusters on tall, hollow stems; the fruits are strongly-winged achenes.

OCCURRENCE Native to Siberia, Rhubarb is widely cultivated in North American gardens for its edible stalk, and also sometimes occurs as an escape. In some areas of the north, where it was planted by early prospectors and settlers, it has become well established. Rhubarb stalks are usually harvested in spring before the plant flowers. Related species are sometimes grown as garden ornamentals.

TOXICITY All parts of the plant, but especially the green leaf blades, contain oxalates of calcium or potassium, and oxalic acid, all of which are irritant poisons. Additionally, anthraquinone glycosides are suspected to be responsible for the highly poisonous nature of the leaves. The fleshy, reddish, sour-tasting leaf stalks are a common garden vegetable, and are safe to eat in moderation. Many cases of poisoning from eating the green leaves of rhubarb have been reported in both humans and animals. This practice was even recommended during World War I in Britain when food was scarce, and some fatalities resulted. The amount eaten to produce symptoms of poisoning may vary from one individual to another. The toxins are not dispelled by cooking.

Symptoms of poisoning may begin within an hour of eating the leaves, and include severe abdominal pains, nausea, vomiting, weakness, difficulty breathing, burning of the mouth and throat, drowsiness, muscular twitching, and in severe cases, convulsions, coma, and death. Even if the patient recovers, liver and kidney damage may occur.

TREATMENT Induce vomiting, or perform gastric lavage; (DO NOT empty stomach if concentrated doses of oxalic acid have been ingested); give lime water, chalk, any soluble calcium salts, milk, or calcium lactate orally to bind the oxalic acid; monitor heart, respiration, and kidney function; maintain fluid and electrolyte balance; give calcium gluconate 10% i.v. to alleviate acute calcium deficiency; furosemide as required to alleviate kidney disfunction from deposition of calcium oxalate crystals; supportive treatment for symptoms.

NOTES There are several other plants in the buckwheat family known to contain oxalates. Some, such as Garden Sorrel (*Rumex acetosa*), Sheep Sorrel (*Rumex acetosella*), Knotweeds (*Polygonum* spp.), and Mountain Sorrel (*Oxyria digyna*) are edible in moderation, and the oxalic acid they contain gives them a pleasant, sour taste. However, if eaten in large quantities or over a prolonged period, they can cause poisoning and interfere with the body's calcium metabolism.

Fig. 162. Snowdrop (*Galanthus nivalis*). R. & N. TURNER

Snowdrop
(Amaryllis Family)

Galanthus nivalis
(Amaryllidaceae)

QUICK CHECK Spring-blooming bulb plant with narrow, elongated leaves and single, drooping, white flowers marked with green; bulb causes digestive upset if eaten, but human fatalities unknown.

DESCRIPTION A herbaceous perennial growing from a small, membranous-coated bulb, often occurring in dense clumps. The leaves, basal and usually 2–3 per bulb, appear with the flowers in early spring and die down after the seed capsules ripen. They are long, narrow, smooth-edged, and somewhat fleshy. The nodding flowers are borne singly at the ends of long stalks. The three outer "petals" are pure white and elongated; the inner ones are shorter, and green-tipped. A double form is frequently grown. The fruit is a green, globular berry containing several seeds.

OCCURRENCE Native to Europe and western Asia, Snowdrop is commonly cultivated throughout North America as one of the earliest blooming spring flowers. Other species of *Galanthus* are also sometimes grown.

TOXICITY The bulb contains toxic alkaloids, including lycorine and galantamine. Eating large quantities can cause nausea, persistent vomiting, and diarrhea, but the plant is apparently not as toxic as its relative, Daffodil, and no fatalities have been reported.

TREATMENT Induce vomiting, or perform gastric lavage; treat for symptoms; maintain fluids and electrolyte balance in cases of heavy vomiting.

NOTES Other plants in the amaryllis family, including *Hippeastrum* spp. (Amaryllis Lilies) and *Crinum*, often grown indoors in winter and outdoors in summer for their large, attractive flowers, may also be dangerous. Their bulbs, like those of Snowdrop, contain lycorine and related alkaloids. However, human poisonings from these plants are not common, because the concentrations of the toxins are relatively low.

Spurges: including Caper Spurge, Snow-on-the-Mountain, Sun Spurge, and Petty Spurge
(Spurge Family)

Euphorbia spp.
(Euphorbiaceae)

Fig. 163. Cushion Spurge (*Euphorbia polychroma*).
R. & N. TURNER

QUICK CHECK Large group of upright or prostrate herbs or shrubs; highly variable in form; most with small, inconspicuous flowers but some with large, showy bracts; most exude a milky juice if cut or bruised; some are toxic and potentially fatal if ingested; sap may irritate eyes and skin.

DESCRIPTION There are over 1600 species in the genus *Euphorbia*. They are extremely variable in form, including herbs, shrubs, trees, and succulent, thorny cactus-like types. Many, including Cushion Spurge (*Euphorbia polychroma;* syn. *E. epithymoides*), Caper Spurge (*E. lathyrus*) and Snow-on-the-Mountain (*E. marginata*), are grown as garden ornamentals. The last is particularly attractive, with conspicuous white margins around its upper and floral bracts. Other spurges, such as Sun Spurge (*E. helioscopia*) and the diminutive Petty Spurge (*E. peplus*), are weedy. Many are grown as potted household ornamentals, or outdoors in warmer regions. Two well known spurges, commonly grown indoors as house and greenhouse plants, are discussed separately in Chapter V: Crown-of-Thorns (*E. milii*) (p. 228), and Poinsettia (*E. pulcherrima*) (p.254). The flowers of most spurges are small, greenish, and inconspicuous, but some (such as Poinsettia, Crown-of-Thorns, Cushion Spurge, and Snow-on-the-Mountain) have large, showy bracts surrounding the flowers. The fruits are 3-valved capsules. Many spurges exude a milky latex when cut or bruised.

OCCURRENCE Spurges are found throughout much of North America, growing as weeds of roadsides and waste land, in gardens, and as household ornamentals. Cushion Spurge and Caper Spurge are

Fig. 164. Snow-on-the-Mountain
(*Euphorbia marginata*). R. & N. TURNER

natives of Europe. Snow-on-the-Mountain is native to the Great Plains, but is planted in flower gardens throughout North America. Sun Spurge and Petty Spurge are common weeds.

TOXICITY Many spurges contain an irritant milky latex or sap, which exudes from the plants when cut or crushed. The toxins of various species have not yet been fully identified. Complex terpenes (including euphorbol) have been implicated, as well as a resin, an alkaloid (euphorbine), a glycoside, a dihydroxycoumarin, and a complex substance named euphorbiosteroid. The toxic properties remain even in dried plants. Human poisoning by spurges has resulted from people mistaking the seed capsules of Caper Spurge for true capers (which are flower buds of an unrelated plant, *Capparis spinosa*). In one instance, two children were poisoned, one of them fatally, from sucking the juice of Sun Spurge. In another case, a young woman was fatally poisoned when she attempted to use Snow-on-the-Mountain as an abortive.

Symptoms of poisoning include intense burning of the mouth, throat, and stomach, salivation, vomiting, convulsions, constriction of the pupils, fluid buildup in the lungs, and, in severe cases, coma and death. Even licking the fingers after handling spurge plants can cause severe irritation of the lips and tongue. Spurges frequently cause rashes and blistering on the skin of people touching them. Eye irritation can also develop if the latex comes in contact with the eyes.

TREATMENT Induce vomiting, or perform gastric lavage; treat for symptoms; maintain fluid and electrolyte balance in cases of severe vomiting. For eye irritation, flush with a gentle stream of tepid water for 15 minutes; for skin irritation, wash skin thoroughly with soap and water.

NOTES Spurges have long been used in folk medicine. The genus was named after Euphorbus, physician to King Juba II of Mauritania in A.D. 18. The main use of the plants has been as purgatives and emetics, but since their action is drastic, they are seldom used today.

Fig. 165. Star-of-Bethlehem (*Ornithogalum umbellatum*). R. & N. TURNER

Star-of-Bethlehem
(Lily Family)

Ornithogalum umbellatum
(Liliaceae)

QUICK CHECK Perennial with onion-like bulbs, grass-like leaves, and star-like, clustered flowers with "petals" white inside, and green and white outside; all parts of the plant, especially the bulbs, are toxic and potentially fatal.

DESCRIPTION A perennial herb growing from a tufted bulb, often occurring in dense clumps. The leaves are narrow and grass-like, with a conspicuous, white mid-vein. The flowers, in clusters of 5 to 20, are star-like and 6-parted, up to 2.5 cm (1 in.) across, the "petals" white inside and green with white outside. The fruits are 3-lobed capsules, each containing a few, dark-colored seeds.

OCCURRENCE A native of Europe and North Africa, Star-of-Bethlehem is frequently grown in North American gardens and can also be found as a weedy garden escape, especially in the southeastern United States. A related white-flowered species, Wonder Flower (*Ornithogalum thyrsoides*), is sometimes grown in North American gardens, and is also poisonous.

TOXICITY All parts of the plant, especially the onion-like bulbs, are poisonous. Lampe and McCann (1985) identify the toxins as con-vallatoxin and convalloside, the same digitalis-like glycosides found in Lily-of-the-Valley. Other sources mention various unidentified alkaloids. Eating the bulbs, leaves, or flowers can lead to pain in the mouth and throat, nausea, vomiting, abdominal pain and cramps, and diarrhea. Slow or irregular heart beat may also occur. Some deaths from children eating this plant have been reported, and livestock may be poisoned from browsing the plants or bulbs exposed from plowing or frost heave. Juice from the plant can irritate the skin.

TREATMENT Induce vomiting, or perform gastric lavage; give repeated doses of activated charcoal; give saline cathartics; closely monitor heartbeat and serum potassium level; atropine or transvenous pacing may be required to regulate heartbeat; phenytoin or lidocaine is recommended for rhythm disturbances (see also under Foxglove—p. 192).

NOTES Several other flowering bulb plants of the lily family are poisonous to some extent. Hyacinth (*Hyacinthus orientalis*), widely grown both as an early spring flower in the garden and as a house plant, has bulbs that are occasionally mistaken for onions and, if eaten in any quantity, can cause stomach cramps, vomiting, and diarrhea. Hyacinth juice can also cause skin irritation. Its bright flowers, up to 2.5 cm (1 in.) across and in colors ranging from blue to red, to white, to yellowish, are usually highly scented, and are arranged in a dense cylindrical cluster along an upright stem.

Another toxic flowering bulb commonly grown in gardens is Bluebell, or Wild Hyacinth (*Endymion hispanicus, E. nonscriptus;* syn. *Scilla* spp.). The flowers are smaller than those of Hyacinth, and are commonly light blue, but also occur in pink and white forms. The bulbs contain glycosides similar in structure to the cardiac glycosides of Digitalis (p. 192). Squill (*Drimia maritima;* syn. *Urginea maritima*) is a related species, sometimes cultivated in California and elsewhere as a garden ornamental or for its medicinal and rodenticide properties. Its bulb contains similar Digitalis-like toxins.

Fig. 166. Bluebell (*Endymion hispanicus*). R. & N. TURNER

Sweet Pea and Its Relatives
(Bean, or Legume Family)

Lathyrus odoratus and related
   species
(Fabaceae, or Leguminosae)

Fig. 167. Sweet Pea (*Lathyrus odoratus*). R. & N. TURNER

QUICK CHECK Herbaceous climbing vine with showy, clustered, pea-like, fragrant flowers in a variety of colors, and narrow, pea-like seedpods; seedlings, pods, and seeds cause paralysis and may be fatal if eaten in large quantities.

DESCRIPTION Herbaceous, annual vine, with hairy, winged or strongly angled stems up to 2 m (6 ft) or more high. The leaves consist of a single pair of elliptic to lance-shaped or oval leaflets, with well-developed, branching tendrils at the tip. The flowers are large, showy, and fragrant, borne in clusters of 2 to 5 on long stalks growing from the leaf axils. Occurring in a wide variety of colors, from red to white to dark purple or blueish, the flowers are typically pea-like, with a large upper petal, or banner, two outer petals at the sides (wings), and a keel of two inner petals joined together. The pea-like pods are narrow, elongated, and hairy, containing 4 to 10 seeds. There are many other *Lathyrus* species, most with pinkish or purple flowers, some white or yellow-flowered. All should be considered inedible, despite their pea-like appearance.

OCCURRENCE This attractive flowering vine, originating from Europe, is widely grown in North American gardens, and its fragrant blooms are popular as cut flowers. It occasionally also occurs as a garden escape. There are many wild relatives of the Sweet Pea, some native and some naturalized from other regions.

TOXICITY Sweet Pea and its relatives contain numerous toxic amino acids called lathyrogens, and their derivatives, notably aminopropionitrile. The highest concentrations are in the seedlings and pods, and, especially, the seeds. Sweet Pea seedlings also contain isoxazolin–5-one derivatives which make them even more poisonous than the dry seeds. Lathyrogens have been responsible for a disease known as "lathyrism," characterized by paralysis, weak heart beat, shallow breathing, muscular tremors, and convulsions in both humans and animals eating the seeds or plants of Sweet Pea or other *Lathyrus* species. Animals fed the seeds have also developed severe skeletal abnormalities, dilation and rupture of arteries, and damage to connective tissue. Fortunately, poisoning occurs only after large quantities of the seeds or plants are eaten; when small quantities are eaten, the effects are said to be negligible.

TREATMENT Few or no symptoms will be evident if small quantities of the seeds are eaten; for larger amounts, induce vomiting, or perform gastric lavage, and give general supportive treatment as required.

NOTES The pea-like seeds of Sweet Pea are commonly eaten by children, and poison control centers are frequently contacted about them. They are apparently not dangerous unless large quantities are eaten, but nevertheless, children should be warned against them (see also under Peas, Wild in Chapter III, p. 96).

Tobacco and Its Relatives
(Nightshade Family)

Nicotiana tabacum and
    related species
(Solanaceae)

Fig. 168. Native Tobacco (*Nicotiana rustica*).
R. & N. TURNER

QUICK CHECK Tobacco is widely known as the commercial smoking substance in cigarettes and cigars and as a crop plant, with is large, bright green leaves. Other species are grown as garden ornamentals, having smaller leaves and trumpet-shaped, petunia-like flowers of various colors: white, yellowish, red, or dark purple. Other tobacco species are wild or naturalized, and one is a small tree. All have tubular, 5-lobed flowers that are flared at the mouth, and most have simple, broad, hairy, strongly scented leaves. All tobaccos should be considered poisonous to eat; some have caused fatalities.

DESCRIPTION Tobacco is a well known crop plant with large, bright-green leaves that are hairy and strong-smelling. There are various other species of *Nicotiana* occurring in North America, some native, some naturalized, and some grown as garden ornamentals. They include

Fig. 169. Tree Tobacco (*N. glauca*). R. & N. TURNER

Fig. 170. Nicotiana (*N. alata*). R. & N. TURNER

annuals, perennials, and even a small tree. In general, the leaves of Tobaccos are large, often 15–30 cm (6 in. to 1 ft) or more long, simple, hairy, and often sticky. The flowers, often strongly fragrant, range from white, yellow, or greenish, to purplish, and usually open at night. The petals are fused into a 5-lobed tube, flared at the mouth. The fruits are dry capsules containing many minute seeds. All tobaccos should be considered poisonous. The most significant toxic species are: Cultivated Tobacco (*Nicotiana tabacum*), Flowering, or Scented Tobacco (*N. longiflora*), Tree Tobacco (*N. glauca*), and Desert Tobacco (*N. trigonophylla*). Two other species commonly found in parts of North America are *N. attenuata* and *N. rustica*.

OCCURRENCE Cultivated Tobacco is grown as a crop plant in many parts of the United States and southern Canada. Flowering Tobacco, also known as Nicotiana, is widely grown as a garden flower and is a common garden escape in some areas. Tree Tobacco, a yellow-flowered shrub or small tree up to 5.5 m (18 ft) tall, is grown as an ornamental, and is also found as an escaped weed in Florida, California, and Hawaii. Desert Tobacco grows as a wild annual in the dry soils of the southwestern United States and Mexico.

TOXICITY Commercial Tobacco contains a highly toxic alkaloid, nicotine, a well known narcotic which is generally consumed in North America through smoking. Poisoning through intentional or accidental misuse of nicotine and products containing it is a relatively common occurrence. It is readily absorbed from ingestion of the leaves, or from inhalation, and in pure form is rapidly fatal in small amounts. Related species may contain other toxic alkaloids, chemically similar to nicotine.

Because of its widespread use, the long-term health risks of

Tobacco have been well studied, and it is implicated as a cause of or contributor to many health problems, including heart disease, cancer of the lungs and throat, and complications in pregnancy and child-birth for women smokers. Aside from these problems, the direct ingestion of raw or cooked leaves of any *Nicotiana* species can cause severe poisoning, even death. Tobacco sometimes causes poisoning when used as a home remedy, or from absorption of nicotine through the skin during commercial tobacco harvesting.

Symptoms of nicotine poisoning are: nausea, severe vomiting, abdominal pain, diarrhea, headache, dizziness, confusion, sweating, salivation, rapid, irregular heart beat; initial high blood pressure followed by low blood pressure; convulsions, collapse, and possible sudden respiratory failure. Although Tobacco plants are distasteful, grazing animals may be poisoned by them under some conditions.

TREATMENT Do not induce vomiting, or perform gastric lavage if patient has vomited or is undergoing seizures; sedate with i.v. diazepam, then administer activated charcoal and a saline cathartic (repeat charcoal and cathartic every 6 hours for 24 hours); strong tea or tannin administered orally has been recommended; oxygen and artificial respiration may be required; maintain fluid and electrolyte balance; monitor heart beat and arterial blood gases; control convulsions with i.v. diazepam or phenobarbital; atropine or phentolamine may be used to control high blood pressure and related symptoms. Recovery from acute phase may be rapid, but residual constipation and urine retention may require treatment.

NOTES Children have been poisoned from sucking the flowers of Tree Tobacco, and the chopped leaves of this species eaten as a salad caused death. In the 1960s, a California family was reportedly poisoned, one member fatally, when they cooked and ate the leaves of Desert Tobacco as a potherb.

This book cannot replace the advice and assistance
of qualified medical personnel.
In all cases of suspected poisoning by plants, or any other substance,
immediate qualified medical advice and assistance should be sought.

Fig. 171. Tulip (*Tulipa* cultivar). R & N. TURNER

Tulip
(Lily Family)

Tulipa spp.
(Liliaceae)

QUICK CHECK Bulb-bearing perennials with smooth, broad, pointed leaves, simple stems and (usually) single, showy, upright, cup-shaped flowers in a wide variety of colors and forms; onion-like bulbs poisonous, but not considered fatal.

DESCRIPTION Spring-flowering herbaceous perennials growing from a tapering bulb with a thin, brown, papery skin. The leaves, usually 3–4 borne on the lower part of the stem, are smooth, somewhat fleshy, broad, oval to lance-shaped or elliptical, and pointed. The flowers grow (usually) singly at the end of an erect stalk, up to 15 cm (6 in.) or more tall. The buds are ovoid and pointed, and the six colored "petals" overlap to form a cup-shaped to saucer-shaped, upright flower. There are innumerable varieties with a wide range of flower colors—scarlet, pink, white, yellow, orange, dark purple, and variegated. Some have a dark purple or blackish center to the flower, some have deeply cut petals, and some have variegated leaves.

OCCURRENCE Tulips are native to the Middle East and Asia. In all there are about 60 species. *Tulipa gesneriana* is the most commonly grown species in North American gardens, and is widely planted in flower borders and beds.

TOXICITY Tulips are known to cause poisoning, and substances named tulipalin A and B have been isolated, but have not definitely been linked with the plant's toxicity. Another compound, acylglucoside tuliposide A, is known to be allergenic. Human poisoning has occurred from eating the bulbs, either by mistaking them for onions, or as a famine food, as occurred in the Netherlands during the Second World War. Recently six members of a Yugoslavian family ate a goulash containing tulip bulbs in place of onions. Within ten minutes, they developed poisoning symptoms, including nausea, vomiting, sweating, increased salivation, difficult breathing, and palpitations. They recovered, but weakness persisted for several days. A dog fed some of the goulash vomited repeatedly and lost its appetite for three days. Handling Tulip bulbs may cause skin irritation, especially among those collecting, sorting, and packing them.

TREATMENT Induce vomiting, or perform gastric lavage; follow with activated charcoal and a saline cathartic; maintain fluid and electrolyte balance; supportive treatment for symptoms.

NOTES Tulip flowers are fed to cattle in the Netherlands, apparently without ill effect, but the bulbs are known to be severely purgative. Apparently there have been no instances of Tulip poisoning in North America, since the plant is scarcely mentioned in North American literature as being toxic.

References for Chapter IV: Bailey 1976; Canadian Pharmaceutical Association 1984; Claus et al. 1970; Cooper and Johnson 1984; Emboden 1979; Fuller and McClintock 1986; Hamon et al. 1986; Hardin and Arena 1974; Hedrick 1972; Hill 1986; Keeler 1979; Kinghorn 1979; Kingsbury 1964, 1967; Lampe and McCann 1985; Levi and Primack 1984; Lewis and Elvin-Lewis 1977; Millspaugh 1974; Morton 1958, 1971; Schultes and Hofmann 1980; Turner 1982, 1984; Tyler 1987.

CHAPTER V

POISONOUS HOUSE PLANTS AND PLANT PRODUCTS
(Including Garden Plants of Subtropical Regions)

This book cannot replace the advice and assistance
of qualified medical personnel.
In all cases of suspected poisoning by plants, or any other substance,
immediate qualified medical advice and assistance should be sought.

WOODY PLANTS

Avocado, or Alligator Pear
(Laurel Family)

Persea americana
(Lauraceae)

Fig. 172. Avocado fruit (*Persea americana*), showing large seed. R. & N. TURNER

QUICK CHECK Small to medium-sized tree with large oval, pale green, often drooping leaves; flowers small, greenish yellow. Large, edible fruits well known and widely marketed, but leaves, bark, unripe fruit, and seeds toxic to animals and potentially to humans. Keep young children away from leaves and seeds.

DESCRIPTION Broad-topped tree up to 20 m (60 ft) high, but much smaller when grown indoors. Leaves 8–15 cm (3–6 in.) long, oblong to oval, pale green and somewhat drooping. Flowers, if they occur, are small and greenish-yellow, in compact clusters. The edible fruits are well known and widely marketed in North America. They are large (5–20 cm, or 2–8 in. long), oval or pear-shaped, and fleshy, with greenish, maroon, or blackish skin, light green, creamy edible flesh, and a single large central seed, rounded at one end, pointed at the other.

OCCURRENCE Avocado is native to Mexico and Central America, and is now grown as a crop in California, Florida, and other subtropical areas. It is often grown indoors as a potted tree, even in northern regions.

TOXICITY Although the ripe flesh of the avocado fruit is edible and widely used by humans, the leaves, unripe fruit, bark, and seeds have been reported to be toxic, sometimes fatally, to various types of animals, including cattle, horses, goats, rabbits, fish, and canaries. For the last, poisoning resulted from eating ripe fruit. The toxins are apparently unknown, but some varieties, including Fuerte and Nabal strains, are more poisonous than others. A major symptom of Avocado poisoning in cattle, goats, and horses is severe mastitis in female animals, with permanent reduction in milk flow. The assumption should be made, unless proven otherwise, that all parts of the plant except the flesh of the ripe fruit may be dangerous for humans.

TREATMENT Treat for symptoms.

NOTES Children often undertake to grow the large, unusual looking seeds, which sprout easily and grow readily into attractive small trees. In the Philippines, Avocado seeds are sometimes used to relieve toothaches.

Castor Bean
(Spurge Family)

Ricinus communis
(Euphorbiaceae)

Fig. 173. Castor Bean (*Ricinus communis*).
R. & N. TURNER

QUICK CHECK Bushy, shrublike, ornamental herb with large, long-stalked, palmately lobed leaves, spiny, clustered seed pods, and bean-like seeds usually mottled with white, red, or brown. Seeds violently poisonous, sometimes fatal, especially for children; symptoms may be delayed for hours or days after ingestion. Castor oil strongly cathartic.

DESCRIPTION Large annual with branching stems up to 3.6 m (12 ft) or more tall, often reddish or purplish tinged. The leaves are large (up to about 80 cm, or 30 in. across), simple, alternate, long-stalked, and palmately lobed, with 5–11 elongated lobes toothed along the margins. The flowers are inconspicuous and lacking petals, but the three-parted fruiting pods are showy, clustered, green or often bright red, and covered with soft spines. The seeds are large and bean-like, glossy black or white, or (usually) mottled black or brown on white.

OCCURRENCE Native to tropical Africa, Castor Bean is naturalized as a weed in warmer parts of the United States, and is grown commercially in California and the southern states for the oil extracted from its seeds. It is often grown as an ornamental, either indoors in cooler climates, or in yards and gardens in warmer areas, because of its striking foliage and rapid growth.

TOXICITY The attractive, usually mottled seeds are the most toxic part of the plant; the leaves are less poisonous. The main toxin, a plant lectin, is a high molecular weight protein (a toxalbumin) called ricin, which is reputed to be one of the most toxic naturally occurring substances. It inhibits protein synthesis in the intestinal wall. Another lectin, called ricinus agglutinin, is known to coagulate and break down red blood cells. Small quantities of the cathartic, castor oil, are also present. Symptoms of poisoning may not appear for several hours, or even for a few days, after the seeds are eaten. Characteristic symptoms are: burning of the mouth and throat, nausea, vomiting, severe stomach pains, diarrhea (sometimes containing blood and mucus), thirst, prostration and shock from massive fluid and electrolyte loss, headache, dizziness, lethargy, impaired vision, possible rapid heart beat, and convulsions. Two to five hours following ingestion, symptoms of retinal hemorrhaging of the eyes, internal hemorrhaging and fluid buildup in the digestive tract and lungs, and deterioration of the liver and kidneys are evident in serious cases. Death from kidney failure may occur up to 12 days after eating the seeds. Eating 1–3 seeds can be fatal to a child, 2–6 to an adult. If swallowed whole, without being chewed, the seeds are unlikely to be harmful, because the hard seed coat prevents the toxin from being released.

Livestock are sometimes fed Castor Bean products, such as husks, that have been detoxified with heat or steam, but if the detoxification procedures are inadequate, severe poisoning and death can result. Cattle, sheep, horses, pigs, and poultry have all been poisoned from meal contaminated with Castor Bean toxins.

TREATMENT Immediately induce vomiting, or perform gastric lavage;

follow with activated charcoal and a saline cathartic; repeat charcoal and cathartic every 6 hours for 24 hours unless severe vomiting and diarrhea ensue; supportive therapy for symptoms; antacids may provide relief; maintain fluid and electrolyte balance; maintain high urine output and keep urine alkaline with 5–15 g (0.2–0.5 oz) sodium bicarbonate daily; monitor blood pressure, electrolytes, heart, kidney, and liver function; blood transfusions may be required.

NOTES Because this plant is commonly cultivated, children often have access to the seeds and sometimes make necklaces and other jewelry from them. They are a major cause of poisoning among children. If the plant is grown as a household ornamental, the seed heads should be removed before the seeds are allowed to mature.

Castor oil is used in medicine as a laxative, and is employed industrially as an ingredient in soaps, varnishes, and paints, and as a lubricant.

Fig. 174. Seed necklace, with large white-and-black mottled Castor Beans. R. & N. TURNER

Croton, or Hogwort
(Spurge Family)

Codiaeum variegatum and
related species (syn. *Croton*
spp.)
(Euphorbiaceae)

Fig. 175. Croton (*Codiaeum variegatum pictum*).
R. & N. TURNER

QUICK CHECK Shrub or small tree with leathery, laurel-like, often lobed, prominently-veined leaves marked with white, yellow, red, or pink. Leaves and stems of many species contain a highly toxic, irritant, and purgative oil; keep young children away from the brightly colored leaves.

DESCRIPTION An attractive shrub or small tree, with oval to lance-shaped, often lobed, leaves that are strongly veined and marked with white, yellow, or reddish coloring, with various forms and combinations of variegation. The flowers are small, in elongated racemes, borne in the axils of the upper leaves. Male and female flowers are on separate plants. The fruit is a globose, 2-parted capsule.

OCCURRENCE Native to the South Pacific and Australia, Crotons are widely cultivated as tropical garden, greenhouse, and house plants for their handsome, variegated leaves.

TOXICITY Many of the Crotons contain croton oil, predominately in the seeds but also in the leaves and stems. The oil is a drastic purgative and is also highly irritating to the skin. Even minute quantities of the pure oil are potentially fatal to humans and domestic animals. Croton oil consists of a mixture of glycerides with component acids, but its irritant, cathartic properties are not due to these, but to an accompanying mixture of terpenoid principles known as phorbols. The seeds of at least some, probably all, Crotons also contain a very toxic albuminous substance, crotin, which is similar to ricin found in Castor Bean. Croton oil is known to act as a secondary cancer-causing agent, or cocarcinogen, activating malignancies in areas previously exposed to a carcinogenic substance. Fortunately Crotons are distasteful and actual cases of poisoning of humans and animals are rare. Nevertheless, because the leaves are so attractive and showy, the ornamental Crotons should be regarded with great caution, especially around young children.

TREATMENT Induce vomiting, or perform gastric lavage; follow with activated charcoal and a saline cathartic; treat for symptoms of skin irritation and gastroenteritis.

NOTES Croton oil was commercially extracted from the seed kernels of *Croton tiglium.* It has been used in medicine, like Castor Oil, as a drastic purgative, but is considered the most violent of all purgatives and one of the most powerful local irritants, as well as having cocarcinogenic properties; hence its use is seldom recommended today.

Crown-of-Thorns
(Spurge Family)

Euphorbia milii (syn.
    E. splendens)
(Euphorbiaceae)

Fig. 176. Crown-of-Thorns (*Euphorbia milii*).
R. & N. TURNER

QUICK CHECK Very spiny, shrub-like plant with a few elongated leaves and flowers borne above 2 bright red bracts; stem and leaves exude milky juice when cut or bruised. Milky sap an irritant of the skin and digestive tract; spines may cause painful injury. Severe human poisoning is rare because the plant is so spiny.

DESCRIPTION A woody, shrublike plant up to 1.2 m (4 ft) high, with stems that are thickly covered with sharp spines up to 2.5 cm (1 in.) long. The leaves are few and scattered, mostly on the young growth, thin and oblong to spoon-shaped, smooth edged, and up to 5 cm (2 in.) long. The "flowers," borne in elongated, hanging clusters, are subtended by two showy red bracts. The fruit is a 3-lobed capsule.

OCCURRENCE Native to Madagascar, this attractive, cactus-like shrub is frequently cultivated as a house and patio plant in North America, and is grown in gardens of the southern United States.

Fig. 177. Candelabra Cactus (*E. lactea*).
R. & N. TURNER

TOXICITY The milky sap, or latex, which exudes from the stem and leaves of this and related plants, consists of complex terpenes, notably euphorbol. It can produce severe irritation of the mouth and digestive tract, and is corrosive and caustic to the skin and eyes, causing temporary blindness in some cases. Fortunately, no severe cases of poisoning by Crown-of-Thorns are on record, probably because it is so prickly that children would not be inclined to eat it. The spines can produce painful injury.

TREATMENT Internal poisoning by this plant is unlikely, but if it occurs, apply general supportive treatment for symptoms, including giving plenty of fluids. If plant or its latex is touched, wash the skin thoroughly with soap and water. Wash eyes with warm saline solution if they are affected.

NOTES According to legend, it was a wreath of this plant that was placed as a mock crown on the head of Christ at his crucifixion. Two related spiny shrubs that are grown as ornamentals in the southern United States and are potentially irritating to the skin and eyes are Candelabra Cactus (*Euphorbia lactea*) and Pencil Tree (*E. tirucalli*).

Common Fig and Its Relatives
(Mulberry Family)

Ficus carica and related species
(Moraceae)

Fig. 178. Common Fig *(Ficus carica)*. R. & N. TURNER

QUICK CHECK Common Fig is a small tree with large, deeply lobed leaves. Other species are broad-leaved shrubs, vines, or trees, often grown in houses, offices, and malls as ornamentals; leaves have various sizes and shapes; stems and leaves exude a milky juice when cut or bruised; fruits and leaves can cause skin irritation, but severity varies with species.

DESCRIPTION Many species of Fig are grown as indoor ornamentals in temperate North America, or outdoors in subtropical areas. The Common Fig *(Ficus carica)* is widely grown in greenhouses or, more commonly, outdoors in warmer areas, for its edible fruit and attractive large, deeply lobed leaves.

There are over 800 species of *Ficus*. They are woody shrubs, vines, or trees native to warmer regions. Some have large, shiny, leathery leaves, some smaller leaves. The leaves of some are deeply or shallowly lobed. Fiddle Leaf Fig *(Ficus lyrata)* has large, oblong, lyre-shaped leaves. All Figs share the characteristic of exuding milky white latex from their leaves and stems if cut or bruised.

OCCURRENCE Figs are grown as pot and patio plants, indoors or outdoors in warmer areas, throughout temperate North America. The Common Fig *(Ficus carica)*, native to the Middle East, is widely grown in greenhouses and conservatories or outdoors in warm, sheltered areas, even in southern Canada. Probably the most common indoor species are the large leaved Indian Rubber Plant *(F. elastica)* and the smaller leaved tree, *F. benjamina*. The latter species is an ideal decorative plant for offices and shopping malls.

TOXICITY One species of Fig in particular, Fiddle Leaf Fig *(F. lyrata)*, was found to be highly toxic to rats in laboratory experiments, and should be considered dangerous for humans until proven otherwise. The leaves, fruits, and sap of several species of Fig are known to cause irritant dermatitis or photodermatitis (skin irritation in the presence of ultaviolet light). The seriousness of the reaction depends on the species or variety of Fig and, apparently, on the person affected. Fig dermatitis, mainly from the Common Fig, can be quite severe, with blistering and redness of the skin and around the mouth of those who harvest, process, or eat fresh Figs or come in contact with their foliage. The blistering and irritation is sometimes followed by discoloration that can last for years. Some Fig species may also cause allergic contact dermatitis, similar to that of Poison Ivy and Poison Oak (*Toxicodendron* spp.), and their sap can also be highly irritating to the eyes. The principal irritating compound, contained in the sap, seems to be mainly ficin, an enzyme causing protein breakdown. Some Figs are also known to contain ficusin, a psoralen, and 8-methoxypsoralen, which are phototoxins.

TREATMENT General treatment for symptoms; thoroughly wash skin and clothing coming in contact with the sap of the leaves, fruit, or peel of any types of Figs. Avoid exposure to sunlight, ultraviolet light, or fluorescent light for at least two weeds after contact. Cold compresses may ease the burning sensation or severe irritation at the acute stage, and aspirin may reduce inflammation.

NOTES In warmer countries where Figs are produced commercially, those who dry, pack, or cook them sometimes develop chronic eczema of the hands. Fig leaves are reputed to have been used by Adam and Eve as their rudimentary clothing, but at least one researcher has pointed out that, in view of their irritant properties, they seem to be most unsuitable for this purpose.

Lantana, or Red Sage
(Vervain Family)

Lantana camara
(Verbenaceae)

Fig. 179. Lantana (*Lantana camara*). R. & N. TURNER

QUICK CHECK Branching shrub with yellow to orange flowers in showy, flat-topped clusters; leaves with pronounced unpleasant odor when crushed. Green, unripe fruits a major cause of poisoning in Florida; sometimes fatal, with symptoms delayed 6 hours or more.

DESCRIPTION Bushy, sometimes sprawling shrub up to 1.2 m (4 ft) high as a potted plant when grown in greenhouses, or up to 6 m (20 ft) high in the tropics. The stems are angled, sometimes with short, hooked prickles. The leaves are simple, opposite or whorled, oval shaped, and

Fig. 180. Lantana, berries. R. & N. TURNER

coarse, with toothed margins, 2.5–12 cm (1–5 in.) long, and stalked. The plant is unpleasantly aromatic when crushed or touched. The flowers are small, tubular, and 4-parted, in attractive, dense, flat-topped clusters up to 5 cm (2 in.) across, and on long stalks. The young flowers at the center are usually yellow, changing to orange, then red as they mature. There are also white and pink-flowered varieties. The fruits are spherical, fleshy, and berry-like, at first green, turning blackish at maturity.

OCCURRENCE A native of dry woods of the southeastern United States, Lantana is widely grown as a house, greenhouse, patio, and hanging basket plant throughout temperate North America; it is a common weed and garden ornamental in tropical and subtropical areas. In greenhouses flowering occurs year-round.

TOXICITY This plant is considered to be one of the chief causes of poisoning in Florida. The green, unripe fruits are the most dangerous. According to Lampe and McCann (1985) the poisonous agent is unknown; other sources mention an atropine-like alkaloid, lantanin and a phototoxic triterpene derivative, lantadene A. The latter is activated only in the presence of sunlight. The ripe fruits are apparently not harmful, but the leaves are known to be fatally poisonous to animals, even in relatively small amounts. Symptoms of poisoning, which may be delayed up to 6 hours after ingestion, include weakness and lethargy, vomiting, diarrhea, difficulty in walking, visual disturbance, and, in severe cases, circulatory collapse and death. Acute symptoms resemble atropine poisoning. Chronic poisoning produces edema, liver degeneration, gastrointestinal lesions, and hemorrhages in some organs.

TREATMENT Induce vomiting, or perform gastric lavage; follow with activated charcoal and a saline cathartic; treat as for atropine poisoning—see under Belladonna (*Atropa belladonna*) (p. 183).

NOTES People wishing to grow this attractive houseplant should remove the flowerheads as soon as the flowers begin dropping, thus preventing the formation of the toxic fruits. These have caused fatal poisoning of children who are attracted to the shiny, green, unripe "berries."

Oleander
(Dogbane Family)

Nerium oleander
(Apocynaceae)

Fig. 181. Oleander (*Nerium oleander*). R. & N. TURNER

QUICK CHECK Tall, evergreen shrub with narrow, pointed, leathery leaves having prominent midribs, and large clusters of showy, pinkish or white flowers. All parts of plant, including flower nectar, highly toxic and potentially fatal.

DESCRIPTION An upright, bushy, evergreen shrub or small tree, compact when grown indoors, up to 8 m (25 ft) tall when growing outside. The leathery leaves are opposite or in whorls of 3, narrow and smooth edged, 10–20 cm (4–8 in.) long, tapering at the base to a short stem. They are lighter colored beneath, with prominent yellowish veins and midrib. Leaves of some varieties are variegated white or yellow. The flowers, growing in loose clusters at the tips of the twigs, are large and showy, ranging in color from white to pink to deep red or purplish. They are funnel-like, with 5 broad or finely cut lobes which spread out and twist to the right; double-flowered varieties are often grown. Fruits are paired, elongated follicles.

OCCURRENCE Native from the Mediterranean region to Japan, Oleander is often grown in temperate North America as a house, greenhouse, or patio tree, and is cultivated as a garden ornamental throughout the

Fig. 182. Oleander, flowers.
R. & N. TURNER

southern United States.

TOXICITY Oleander is extremely poisonous. All parts of it, smoke from burning it, and water in which flowers have been placed are toxic. A single leaf is potentially lethal for humans, and eating as little as 0.005 % of the animal's weight is can be fatal to cattle and other animals. People have been poisoned from using Oleander sticks for roasting frankfurters or other meat, and, children in particular, from chewing the leaves or sucking the flower nectar. Honey made from the flower nectar is also poisonous.

The toxins are cardioactive glycosides, especially oleandroside and nerioside, whose action is similar to that of digitalis. Symptoms of poisoning are: nausea, severe vomiting, stomach pain, dizziness, slowed pulse, irregular heartbeat, marked dilation of the pupils, bloody diarrhea, drowsiness, coma, and sometimes paralysis of respiration and death.

TREATMENT Induce vomiting, or perform gastric lavage, followed by repeated doses of activated charcoal and a saline cathartic; supportive therapy for symptoms; monitoring of heart and measurement of serum potassium should be carried out frequently; potassium, procainamide, quinidine sulfate, disodium salt of edetate, or, preferably, dipotassium salt of edetate have all been used effectively.

NOTES A closely related species, *Nerium indicum,* is also sometimes grown in warm climates, especially in Hawaii, and is also potentially fatally toxic. Yellow Oleander (*Thevetia peruviana*), another relative of tropical and subtropical areas, is a shrub or small tree with bright yellow or orange flowers and fleshy fruits, black when ripe and with a central stone. All parts of the plant are dangerous; on the Hawaiian Islands, it is considered the most frequent cause of severe poisoning in humans.

Fig. 183. Poinciana (*Poinciana gilliesii*).
MARY W. FERGUSON

Poinciana and Its Relatives
(Bean, or Legume Family)

Poinciana gilliesii (syn.
Caesalpinia gilliesii) and
related species
(Fabaceae, or Leguminosae)

QUICK CHECK Showy, non-spiny shrub with alternate leaves finely divided into numerous small leaflets; terminally clustered, light-yellow flowers with long, red, conspicuous stamens. Bean-like seed pods may be attractive to children and cause serious irritation of the digestive tract, but are not known to be fatal to humans.

DESCRIPTION An upright, or sometimes climbing, shrub or small tree up to 4.5 m (15 ft) tall. Leaves alternate, and pinnately divided into many small leaflets which are dotted with black near the margins of the lower surface. The flowers are in terminal clusters and extremely showy. They are large and light-yellow, with striking red stamens protruding 8–12 cm (3–5 in.) from the flowers. The fruits are bean-like pods up to 2 cm (0.75 in.) wide and 10 cm (4 in.) long. Several related species (including those classed in the genera *Delonix* and *Caesalpinia*) are grown as ornamental shrubs and trees in the southern United States. Most have feathery leaves and showy flowers with colors ranging from scarlet to orange-yellow to yellow.

OCCURRENCE Poinciana is native to Argentina in South America, and is grown as a large pot plant in houses and greenhouses of temperate North America, and as an outdoor ornamental in subtropical areas.

TOXICITY The green seed pods contain tannins and are severely irritating to the digestive tract of both humans and animals. Two children who each ate 5 pods developed within 30 minutes symptoms of nausea, vomiting, and diarrhea. Fortunately, recovery occurred after about 24 hours.

TREATMENT In severe cases, induce vomiting, or perform gastric lavage; general supportive treatment for symptoms.

NOTES Another, related plant called False Poinciana, Rattlebox, or Sesbane (*Daubentonia punicea*) is sometimes grown as an ornamental shrub and also occurs as a garden escape in the southeastern United States. Its seeds and flowers contain toxic saponins and are potentially fatal.

Poinciana is also sometimes called Bird-of-Paradise, but this name is potentially confusing because it is also applied to a plant in the banana family (*Strelitzia*), which is likewise grown as a large potted plant. It has large, simple, upright leaves with long stalks, and large, showy flowers, and is considered nonpoisonous.

HERBACEOUS PLANTS AND HOUSEHOLD PLANT PRODUCTS

**Alocasia, or
Giant Elephant's Ears**
(Arum Family)

Alocasia spp.
(Araceae)

Fig. 184. Alocasia (*Alocasia* sp.). J. C. RAULSTON

QUICK CHECK Perennial herbs with large, long-stalked, arrowhead-shaped leaves usually marked or colored; entire plant contains irritant calcium oxalate crystals, causing intense burning of the mouth and throat if swallowed; not usually fatal.

DESCRIPTION Perennial herbs growing from rhizomes, with large, long-stalked, entire, scalloped, or lobed, arrowhead-shaped leaves that are often beautifully marked or colored. The flower stalks are shorter than the leaf stalks, and the flowering heads consist of a short, greenish, sheathing spathe subtending a spadix covered with small, unisexual flowers and terminating in a sterile appendage. The most commonly grown species include: Chinese Taro (*Alocasia cuculata*), with large, glossy green leaf blades on stalks up to 1 m (3 ft) long; Kris Plant (*A. sanderiana*), with metallic green, scalloped-edged leaves; Amazon Alocasia (*A. amazonica*), with dark green leaves contrasting sharply with bold, white veins; and Giant Alocasia (*A. cuprea*) with large leaves that are iridescent purple on the upper surface and red-violet beneath.

OCCURRENCE Alocasias are not commonly grown, but they are strikingly spectacular, highly visible plants that are sometimes found in North

American houses, greenhouses, and conservatories. There are about 70 species altogether, most native of tropical Asia, and some of the New World.

TOXICITY Alocasias, like other members of the arum family, contain microscopic, sharp crystals, or rhaphides, of calcium oxalate. If the leaves or any part of the plant are chewed or swallowed, they cause painful irritation and swelling of the mouth and throat. They can also cause skin and eye irritation on contact. For a detailed description of symptoms, see under Anthurium (p. 238).

TREATMENT See under Anthurium (p. 238); check for obstruction of air passage in cases of ingestion.

NOTES The fleshy underground stems and leaf stalks of some *Alocasia* species are a valuable food in parts of the tropics. They are apparently specially prepared to reduce or eliminate the irritant calcium oxalate crystals.

Aloes
(Lily Family)

Aloe spp.
(Liliaceae)

Fig. 185. Aloe Vera (*Aloe barbadensis*). R. & N. TURNER

QUICK CHECK Perennial herbs with fleshy, elongated, often spiny-edged leaves in basal clusters; the juice is strongly purgative if ingested, but not fatal.

DESCRIPTION There are about 150 species of *Aloe,* many of them grown as house and greenhouse plants in North America. They are perennial herbs with basal clusters, or rosettes, of succulent, elongated leaves which are often sharp-pointed, spiny, or hard-toothed on the edges. The flowers, borne in (usually dense) elongated or umbrella-like clusters on thin stems above the leaves, are tubular, and red, pinkish, whitish, or yellow. The fruits are 3-angled capsules. Aloe Vera, or Barbados Aloe (*Aloe barbadensis*), is probably the most commonly grown species in North America. Its erect or spreading leaves are long, narrow, and pointed, edged with short spines. They are glaucous-green with white spots or markings, thick and fleshy, exuding a mucilaginous gel when cut or broken.

OCCURRENCE Most Aloes are native to tropical Africa. Many species are grown in greenhouses in collections of succulent plants, and they are also grown as potted house plants. In the southern United States they are grown outdoors as garden ornamentals. Aloe Vera is native to northern Africa but was introduced into the Barbados Islands in the 17th century; hence the name *Aloe barbadensis.*

TOXICITY The latex in Aloe leaves is poisonous if ingested; it contains a number of anthraquinone glycosides, the principal one known as bar-baloin. The quality and quantity of the glycosides varies from one species to another; Aloe Vera contains relatively high concentrations, and also contains an appreciable amount of chrysophanic acid. These compounds have a strong purgative action, evident 6–12 hours after ingestion, due to irritation of the large intestine. Excessive doses may cause kidney irritation. The anthraquinone compounds color alkaline urine red.

TREATMENT Treat for symptoms; maintain fluids.

NOTES Used for centuries as a skin salve in the treatment of burns, abrasions, and other skin irritations in its native regions, Aloe Vera is now a popular home remedy in North America. It is also grown commercially for its cathartic glycosides, and for its mucilaginous gel, which is found in many household products such as skin lotions, creams, and shampoos.

Anthuriums or Tail Flowers, including Oilcloth Flower and Flamingo Flower
(Arum Family)

Anthurium andreanum,
   A. scherzerianum, and related
   species
(Araceae)

Fig. 186. Anthurium, or Flamingo Flower (*Anthurium scherzerianum*). R. & N. TURNER

QUICK CHECK Large plants with heart-shaped or elliptical leaves and showy "flowers" with a scarlet sheath surrounding an elongated spike; entire plant contains calcium oxalate crystals which irritate the mouth and throat if swallowed; may also irritate skin and eyes.

DESCRIPTION Herbaceous perennials with clustering, dark green, leathery leaves that are heart-shaped in Oilcloth Flower (*Anthurium andreanum*), and lance-shaped in the smaller, more commonly grown Flamingo Flower (*A. scherzerianum*). The leaf blades are up to 30 cm (1 ft) long, and the plants up to 60 cm (2 ft) high. Each showy flower head consists of a shiny, scarlet, reflexed sheath, or spathe, surrounding a red, yellowish, or greenish, club-like spike, or spadix, which is densely covered with small, bisexual flowers and, in Flamingo Flower, is somewhat curled or spiralled. The fruits are densely clustered, showy berries.

OCCURRENCE There are more than 600 species of *Anthurium*, native to tropical America. Anthuriums, especially the two species mentioned, are grown as house and greenhouse ornamentals in temperate North America, or outdoors in the southern United States. Their exotic flower heads last up to two months, and they are commonly sold by florists as cut flowers.

TOXICITY The leaves and stems contain minute, sharp crystal bundles, or raphides, of calcium oxalate, like those found in *Dieffenbachia* and other plants in the arum family. If ingested, the plant causes painful burning and swelling of the lips, mouth, and throat, sometimes with acute inflammation that can restrict breathing. Symptoms may include intense salivation, hoarseness, difficulty in swallowing, loss of speech, and loss of appetite. The initial pain from ingestion almost always prevents further swallowing. Also, the crystals pass unchanged through the digestive tract and so do not usually cause further complications. The calcium oxalate crystals may cause irritation to the skin and eyes. The symptoms may persist for several hours or days.

TREATMENT Give milk or ice cream to sooth irritated mouth and throat; sucking on ice chips or a popsicle may bring relief from intense pain; emptying the stomach is probably not necessary; observe closely for swelling of the tongue and throat that could obstruct the upper air passage. Antihistamines such as diphenhydramine may provide relief for severe swelling, and analgesics may be needed for pain. For

Fig. 187. Anthurium, or Oilcloth Flower (*A. andreanum*). R. & N. TURNER

eye irritation, flush with gentle stream of tepid water for 15 minutes; for skin irritation, wash skin thoroughly with soap and water.

NOTES Many plants in this family, including a number commonly grown as house plants and described in this section, contain irritant calcium oxalate crystals. Many of these are popular ornamentals, often seen in hospital waiting rooms, pre-schools, doctors' offices, restaurants, and shopping malls.

Arum, or Cuckoo Pint
(Arum Family)

Arum spp.
(Araceae)

Fig. 188. Arum (*Arum italicum*), fruiting stage.
RICHARD J. HEBDA

QUICK CHECK Large-leaved herbaceous perennials with attractive but bad smelling "flowers" consisting of a broad, yellowish, whitish or purplish sheath surrounding a cylindrical spike; bright reddish berries are attractive to children; entire plants, especially berries, toxic and potentially fatal if ingested in quantity.

DESCRIPTION Stemless plants growing from tuberous roots, with large, long-stemmed oval to triangular or arrow-shaped leaves. The showy but putrid-smelling flower heads are long-stalked, and consist of a broad sheath, or spathe, which is greenish, yellowish, cream-colored, or violet, enclosing a dull purple to brownish, cylindrical spike, or spadix. This spadix is crowded with tiny, unisexual flowers, the male flowers situated above the female. The flowering stalk tends to wither rather than fall after maturity, but one species, *Arum pictum*, blooms in fall. The fruits form a dense cluster of fleshy berries, green at first and ripening to reddish orange or brilliant red after the sheath and leaves die down.

OCCURRENCE Arums originated in Europe and the Middle East, and a few species are grown for curiosity as house plants in North America. They are sometimes grown outdoors in the southern United States and in mild climates further north. One commonly grown species is *Arum italicum*, some forms of which have cream-veined foliage. Another is *A. maculatum*, known as Cuckoo Pint or Lords and Ladies in Britain, where it grows in the wild.

TOXICITY These plants, like others in the arum family, contain microscopic, sharp crystals, or raphides, of calcium oxalate. These may irritate the skin and eyes, and, if ingested, the lips, mouth, and throat. Arums also contain a volatile, unstable irritant called aroin (a saponic glycoside), which ultimately affects the central nervous system. The attractive berries are particularly poisonous, and are sometimes eaten by children. The acrid juice of the plant usually deters people from eating large amounts, but if sufficient quantities are ingested, it can cause sore throat, digestive irritation and pain, with severe diarrhea, irregular heartbeat, and, in extreme cases, coma and death. Fortunately, poisoning is not usually severe. The plant can irritate the eyes and skin on contact. It is also known to be toxic, sometimes fatally, to various types of livestock, but it is seldom willingly eaten by animals because it is so acrid.

TREATMENT In serious cases induce vomiting, or perform gastric lavage. For further information, see under Anthurium (*Anthurium* spp.) (p. 238); check for obstruction of air passage through swelling in cases of ingestion.

NOTES In England, the rootstock of Cuckoo Pint (*Arum maculatum*), was formerly used as a source of laundry starch, but this practice was abandoned because it caused chronic irritation of the hands. Food preparations of the baked and powdered root were called Portland arrowroot and Portland sago. It is said that the alternate name for this plant, Lords and Ladies, originated in Tudor times from the practice of stiffening courtiers' ruffs with the starchy tubers.

Caladiums, or Angel's Wings
(Arum Family)

Caladium bicolor, C. hortulanum,
   and related species
(Araceae)

Fig. 189. Caladium (*Caladium* hybrid). R. & N. TURNER

QUICK CHECK Perennial herbs with showy, variegated, arrowhead-shaped leaves of a wide variety of colors; entire plant contains irritant calcium oxalate crystals, causing intense burning of the mouth and throat if swallowed; not usually fatal.

DESCRIPTION Perennials growing from underground tubers. The leaves are basal and long-stalked, arising in a clump from the tubers. The blades are large and arrowhead-like, and are veined, edged, or splashed with a wide range of very attractive colors. The "flowers" consist of a dense, fleshy spike, or spadix, subtended by a boat-shaped, often showy bract, or spathe. The berries are white. There are about a dozen species of Caladium, with many popular hybrids.

OCCURRENCE Caladiums are native to tropical America, and are commonly grown as ornamental house and greenhouse plants throughout temperate North America, as well as being cultivated in gardens in subtropical areas.

TOXICITY These plants, like other members of the arum family, contain microscopic, sharp crystal bundles, or rhaphides, of calcium oxalate, which, if swallowed, cause intense irritation and swelling of the mouth and throat, and can also cause skin and eye irritation on contact. For a detailed description of symptoms, see under Anthurium (p. 238).

TREATMENT See under Anthurium (p. 238); check for obstruction of air passage in cases of ingestion.

NOTES Most of the Caladiums sold as house plants in North America are hybrids derived from two species: *C. bicolor* with very colorful, often bright red leaves (sometimes called Heart of Jesus), and *C. hortulanum,* having white leaves with green veins. There are scores of varieties available, but most are sold unnamed. The plants are perennials, but the foliage dies down in the winter.

Calla Lily
(Arum Family)

Zantedeschia aethiopica and
related species
(Araceae)

Fig. 190. Calla Lily (*Zantedeschia aethiopica*).
R. & N. TURNER

QUICK CHECK Large herb with smooth, arrowhead-shaped leaves and showy white-sheathed "flowers"; entire plant contains irritant calcium oxalate crystals which cause intense burning and inflammation of the mouth and throat if ingested; seldom fatal.

DESCRIPTION Calla Lily is a robust, rhizomatous perennial herb up to 1 m (3 ft) high, with smooth-edged, shining, arrowhead-shaped leaves which are basal, clustered, and long-stalked. The "flowers" consist of a showy, flaring white spathe 13–25 cm (5–10 in.) long surrounding a yellow, fragrant, club-like spike of tiny true flowers. The fruit is berry-like. Other species are sometimes also grown, and there are many hybrids in various colors.

OCCURRENCE Calla Lily (not the true Calla; see Notes) is native to South Africa, and is grown as a large, potted house and greenhouse plant throughout North America, and as an outdoor garden plant in the southern United States. It is also found as a garden escape in many frost-free areas of the southern United States.

TOXICITY Calla Lily, like other members of the arum family, contains irritant crystals of calcium oxalate, and if ingested, causes intense burning and swelling of the mouth and throat. It can also cause skin and eye irritation on contact. An unidentified toxic protein has also been reported. For a detailed description of symptoms, see under Anthurium (p. 238).

TREATMENT See under Anthurium (p. 238); check for obstruction of air passage in cases of ingestion.

NOTES Calla Lily is sometimes confused with true Calla—Wild Calla, or Water Arum (*Calla palustris*)—which is a plant of the same family growing in wet, boggy areas throughout much of the United States and Canada and sometimes planted around ponds in gardens.

Fig. 191. Chili Pepper (*Capsicum annuum*).
R. & N. TURNER

Chili Peppers
(Nightshade Family)

Capsicum annuum and
related species
(Solanaceae)

QUICK CHECK Perennial herbs with smooth-edged, elliptical or oval leaves, whitish flowers, and smooth, rounded or elongated fruits that are usually bright red (sometimes yellow or purple) when mature; the fruits contain varying amounts of an irritant principle that can cause painful inflammation and burning of the mouth, mucus membranes, skin, and eyes.

DESCRIPTION Perennial herbs with simple, alternate, elliptical or oval, somewhat thin leaves. The small, star-shaped flowers are borne in the branch axils, and are whitish, usually with a purplish tinge. The fruits are pod-like and many-seeded, with a smooth, fleshy skin. They vary in form, and color, being elongated, pointed or rounded, upright or drooping, and variously colored (white, green, yellowish, or purplish) during ripening, usually bright red (sometimes yellow, orange, or purple) at maturity. Some varieties of *Capsicum* have sweet-tasting fruits; these are the common green, red, and yellow sweet peppers used in salads and cooking in North American households. Others, mostly the smaller, elongated types, are the Chili Peppers. They are hot and biting to the taste and are used as flavoring, particularly in Mexican dishes. Paprika is from yet another variety of *C. annuum*. The pepper used in tabasco sauce is from a related species, Tabasco Pepper (*C. frutescens*).

OCCURRENCE The ancestor to Chili Peppers was probably native to Central and South America. Chili Peppers are cultivated in the southern United States and Mexico for their hot, spicy fruits; some forms grow wild in the South. Many varieties are grown as potted ornamental house plants.

TOXICITY The fruits and seeds of Chili Peppers, especially some of the more pungent varieties, contain the compound capsaicin. It can cause painful, but usually harmless, irritation of the lips and mouth if ingested. If large amounts are eaten, Chili Peppers may cause digestive upset and diarrhea. They may also cause painful inflammation and burning of the mucus membranes and skin on contact. A major danger is if the acrid juice from the fruits gets into the eyes; it can cause acute inflammation, and pain, with heavy tear production. This can happen simply from inadvertant rubbing of the eye with the hands if one has been handling Chili Peppers in cooking. Smoke from cooking the peppers may also irritate the eyes.

TREATMENT For eye contamination with Chili Pepper, flush with a gentle stream of tepid water for 15 minutes; if irritation persists, see an ophthamologist. For ingestion, use demulcents such as milk or ice cream to relieve the burning; sucking on ice or a popsicle may bring relief. Topical anesthetics may be required for severe pain. For skin irritation, wash affected area well with soap and water.

NOTES Chili Peppers are sometimes used as ingredients in tear gas, according to the Canadian Pharmaceutical Association (1984). Fuller and McClintock (1986) report that the irritant smoke from burning Chili Peppers was once used as a form of torture.

Dieffenbachia, or Dumbcane
(Arum Family)

Dieffenbachia spp.
(Araceae)

Fig. 192. Dieffenbachia (*Dieffenbachia amoena*).
R. & N. TURNER

QUICK CHECK Tall herbaceous perennials with large, usually mottled leaves; entire plant contains irritating calcium oxalate crystals, painful to the mouth and throat if eaten, but seldom fatal.

DESCRIPTION Perennial herbs 1–2 m (3–6 ft) tall, with green, unbranched stems having conspicuous, evenly spaced, horizontal leaf scars. The leaves are large, oblong, and smooth-edged, with clasping or sheathing stalks. The veins are prominent and the blades are usually blotched with whitish, light green, dark green, or yellow-green markings. The calla-like flowers, when they occur, are of a typical arum form, with a greenish or whitish sheath surrounding a spadix.

Fig. 193. Dieffenbachia (*D. picta*). R. & N. TURNER

Several species and their cultivated varieties are grown as indoor ornamentals in North America. One of the largest is *D. amoena*, with dark green leaves up to 45 cm (18 in.) long, striped with white bars. Another, smaller leaved type which is very popular is *D. picta.* The leaves of this species range from nearly all green to practially all cream except for a green margin. The plants are grown primarily for their decorative leaves.

OCCURRENCE Native to tropical America, Dieffenbachias are widely grown in temperate regions in houses, hotel lobbies and restaurants, especially areas of low light intensity. They are often found as ornamentals in shopping malls, pre-schools, and health care centers, where their toxic, irritant properties are apparently not well known. In the subtropics they are grown outdoors in gardens.

TOXICITY Dieffenbachias contain microscopic needle-like crystal bundles, or rhaphides, of calcium oxalate, which penetrate the tissues of the mouth and throat if the plant is eaten. Combined with other chemicals, primarily a protein enzyme and asparagine, the crystals cause immediate, severe burning and inflammation of the throat and mouth, and copious salivation, apparently due partly to mechanical and partly to chemical irritation. In severe cases, swelling of the throat and tongue may cause choking (which has occasionally been fatal), and temporary loss of speech (hence the second name, Dumbcane). Nausea, vomiting and diarrhea are additional symptoms, perhaps due to the presence of other, unknown toxins. Symptoms may persist for a week or more. These plants also cause irritation to the skin and eyes.

TREATMENT Emptying the stomach is usually not required unless very large quantities have been ingested; give demulcents (milk, ice-cream, oil, cooling drinks) to soothe irritated membranes; rinse out mouth, and apply ice or a popsicle to lips and mouth to relieve pain; observe patient carefully for symptoms of swelling of the throat and choking. Antihistamines such as diphenhydramine may provide relief for severe swelling, and analgesics may be needed for pain. For eye irritation, flush with gentle stream of tepid water for 15 minutes; for skin irritation, wash skin thoroughly with soap and water.

NOTES Other plants in the arum family with similar toxic components and poisonous effects include the Western yellow Skunk-Cabbage, or Swamp Lantern (*Lysichitum americanum*), the Eastern Skunk-Cabbage (*Symplocarpus foetidus*), the eastern Jack-in-the-Pulpit (*Arisaema* spp.) (p. 115), and several genera of tropical plants that are often used as house plants: Philodendrons (*Philodendron* spp.) (p. 252), Elephant's Ears (*Colocasia* spp.) (p. 247), Alocasias (*Alocasia* spp.) (p. 236), Arums (*Arum* spp.) (p. 240), Caladiums (*Caladium* spp.) (p. 242), Calla Lily (*Zantedeschia aethiopica* and related species) (p. 243), and Anthuriums (*Anthurium* spp.) (p. 238).

Elephant's Ears, or Taro
(Arum Family)

Colocasia esculenta and
    related species
(Araceae)

Fig. 194. Elephant's Ears, or Taro (*Colocasia esculenta*). R. & N. TURNER

QUICK CHECK Large, ornamental, tuberous perennials with huge, shield-shaped leaves; entire plant contains irritant calcium oxalate crystals which cause intense burning to mouth and throat if ingested, but are seldom life-threatening.

DESCRIPTION Large plants growing from turnip-like tubers. The long-stalked leaves are large and shield-shaped, bright green and waxy-looking, forming clumps or patches. The arum-like flowers are apparently seldom seen, and the plants are usually propagated vegetatively.

OCCURRENCE Native to the South Pacific, *Colocasia* species are grown out-doors in the southern United States as ornamentals. In other parts of North America they are used in indoor planters. Taro (*C. esculenta*) is cultivated as a root crop in Hawaii and the South Pacific.

TOXICITY Like other members of the arum family, these plants contain irritant crystals of calcium oxalate, and if ingested, cause intense burning and swelling of the mouth and throat. They can also cause skin and eye irritation on contact. For a detailed description of symptoms, see under Anthurium (p. 238).

TREATMENT See under Anthurium (p. 238); check for obstruction of air passage in cases of ingestion.

NOTES The best known member of this genus, Taro, is a popular root vegetable in the tropics. Its thick, turnip-like rootstocks must be specially prepared to reduce their irritant effect. They are best known as a major ingredient of *poi*, a favourite Polynesian dish. The starch granules of Taro are very small, and reputed to be more digestible than those of other root crops. The leaves and young shoots of some varieties of Taro are eaten as cooked greens.

Jerusalem Cherry
(Nightshade Family)

Solanum pseudocapsicum
(Solanaceae)

Fig. 195. Jerusalem Cherry (*Solanum pseudo-capsicum*). R. & N. TURNER

QUICK CHECK An evergreen shrub with dark green leaves, white flowers, and attractive, round, yellow or red fruits; entire plant, and especially the fruits, toxic and potentially fatal.

DESCRIPTION Bushy, erect, evergreen shrub up to 1.2 m (4 ft) tall, with shiny, dark green, oblong, lance-shaped leaves up to 10 cm (4 in.) long. The flowers are white and solitary, about 1.2 cm (0.5 in.) across. The cherry-sized berries are bright red or yellow, and long-lasting, making the plant attractive and highly ornamental.

OCCURRENCE A native of the Old World, Jerusalem Cherry is commonly grown as a potted house plant, especially during the Christmas season. It is often purchased in nurseries when the berries are green, so their ripening can be enjoyed. It is also grown as an outdoor patio plant, or year-round garden plant in warmer regions. Occasionally it occurs as a garden escape.

TOXICITY The entire plant contains a toxic steroid alkaloid, solano-capsine, which acts on the heart. It also contains the bitter glycoalkaloid solanine and related alkaloids, as are found in other *Solanum* species, including Nightshades (p. 94, 125) and the green tubers and sprouts of Potato (p. 208). The attractive, brightly colored fruits are especially dangerous, and eating 3–4 has proven fatal to children. Symptoms of poisoning, which may be delayed by several hours after ingestion, include scratchy feeling in the throat, abdominal pain, fever, nausea, vomiting, salivation, severe diarrhea, and headache. In severe cases, lowered or irregular heart rate, difficult breathing, restlessness, confusion, delirium, hallucinations, convulsions, kidney failure, and coma followed by death may occur. Recovery may be slow, with symptoms persisting for several days.

TREATMENT Induce vomiting, or perform gastric lavage; follow with activated charcoal and saline cathartic; maintain fluid and electrolyte balance; monitor breathing, heart beat, blood pressure, and kidney function; general supportive treatment for symptoms; antacids may relieve digestive upset; sponge with tepid water to reduce fever; convulsions may be controlled with diazepam i.v.; physostigmine (slow i.v. injection; 2 mg for adults, 0.5 mg for children, repeated 2–3 times as required) is suggested to reverse heart irregularities, convulsions, and hallucinations.

NOTES A related species, known as False Jerusalem Cherry, or Christmas Orange (*S. capsicastrum*), is also a common house plant during the Christmas season. A native of Brazil, it is similar to Jerusalem Cherry, but has smaller berries. Several varieties are grown, including a variegated form with cream-splashed leaves. Another relative, Chinese Lantern (*Physalis alkekengi*), is often found in dried flower arrangements because of its bright red or orange, papery fruiting calyces. Its berries, enclosed within the "lanterns," contain solanine when unripe, and may cause digestive upset if eaten. The ripe berries of this and related species are non toxic and are sometimes made into jams and preserves.

Mistletoe
(Mistletoe Family)

Phoradendron spp. and related
   genera
(Loranthaceae, or Viscaceae)

Fig. 196. Mistletoe (*Phoradendron serotinum*).
WALTER H. LEWIS

QUICK CHECK Partially parasitic plants growing in dense clusters on deciduous and evergreen trees; widely used as traditional household Christmas decorations; berries and green plants produce severe digestive upset when ingested in quantity; rarely fatal.

DESCRIPTION Partially parasitic plants growing on deciduous and evergreen trees, both flowering and cone-bearing. The mistletoes of Christmas are mainly *Phoradendron* species, especially *P. serotinum*. They have smooth-edged, oblong, leathery, green leaves in opposite pairs. The tiny, inconspicuous flowers ripen into clusters of showy, whitish or pinkish, translucent berries. The seeds are sticky and are often spread from one tree to the next by birds. Dwarf Mistletoes (*Arceuthobium* spp.) are smaller, with broom-like aerial shoots, and are parasitic on coniferous trees. They are a major forest pest, responsible for extensive damage in wood production in North America.

OCCURRENCE *Phoradendron* Mistletoes grow in large, dense clusters on deciduous trees. Various species occur from Oregon and California east to New Jersey and Florida. In their fruiting stage, Mistletoes are harvested and distributed commercially throughout North America for use in Christmas festivities. The European Mistletoe (*Viscum album*) has been introduced locally in some areas.

TOXICITY The entire plants, especially the leaves and stems, contain toxic amines and protein substances called lectins, or toxalbumins (mainly phoratoxin), which inhibit protein synthesis in the intestinal wall. Eating only a few berries may cause minor abdominal pain and diarrhea, but ingesting large quantities of the berries, or drinking tea made from the leaves, can produce severe irritation of the digestive tract, including vomiting, diarrhea, and acute cramping. Lowered heart rate, similar to but less severe than that produced by digitalis, may also occur. Mistletoe has on rare occasions been fatal.

TREATMENT Induce vomiting, or perform gastric lavage; follow with activated charcoal and a saline cathartic; treat for severe gastroenteritis; maintain fluid and electrolyte balance.

NOTES An old Christmas tradition, its origins probably extending back to early European ceremonial use, is that any person caught standing beneath mistletoe must forfeit a kiss. Nowadays, to prevent poisoning in children, and to increase the shelf life of Mistletoe in marketing, artificial plastic berries are often substituted for real ones by commercial florists.

Nutmeg
(Nutmeg Family)

Myristica fragrans
(Myristicaceae)

Fig. 197. Nutmeg (*Myristica fragrans*). R. & N. TURNER

QUICK CHECK Brown, ovoid, spicy seed, or grated brown powder, widely used as a culinary spice; nutmeg is harmless in small amounts, but if ingested in larger doses (i.e., greater than 10 g, or 0.4 oz) it can cause acute poisoning and death.

DESCRIPTION Nutmeg is the large, woody, brown seed of a tropical evergreen tree, which yields two spices commonly found in North American homes: nutmeg, and mace, the thin, net-like, reddish or yellowish covering of the seed. These spices can be purchased and stored whole, to be grated or ground as required, or bought and stored in ground form. The Nutmeg tree is tall, with alternate, yellowish brown leaves that are elliptical or lanceolate, and up to 13 cm (5 in.) long. The flowers are small and inconspicuous, male and female borne on separate individuals. The fruit is globular, about the size of a golf ball, reddish or yellowish, and splitting into two valves, each containing a nutmeg enclosed in a hard shell.

OCCURRENCE Whole or ground nutmeg is often found on the spice shelf of North American homes. Occasionally Nutmeg trees are grown in conservatories and botanical gardens in North America. Nutmeg is native to India, Australia, and the South Pacific, and there are about 80 different species in the genus.

TOXICITY Nutmeg contains volatile oils which give it its spicy scent and flavor, but are irritating to all tissues; if excessive amounts are inhaled or ingested, it can cause acute poisoning. The primary constituents of the aromatic ether fraction of nutmeg are myristicin, elemicin, and safrole; these are apparently the toxic principles. In large quantities (over 10 g, or 0.4 oz), powdered nutmeg or mace affects the central nervous system, and produces hallucinations and unpleasant side effects, including headache, dizziness, drowsiness, nausea, stomach pain, excessive thirst, rapid pulse, delirium, anxiety, double vision, and sometimes acute panic and coma. Most cases of nutmeg poisoning occur when people try to use it as a readily obtainable hallucinogenic drug. People who have taken it agree that it causes an agonizing hangover.

TREATMENT Mineral or castor oil, followed by gastric lavage and demulcents is the treatment recommended by Hardin and Arena (1974).

NOTES Many other common household herbs and spices, including cinnamon, cloves, eucalyptus, mint, black pepper, rosemary, sage, and sassafras, contain volatile oils which could be harmful in large or concentrated doses (Hall 1973). Moderation is the key to using these substances.

Philodendrons
(Arum Family)

Philodendron spp. and
 Monstera spp.
(Araceae)

Fig. 198. Heart-leaf Philodendron (*Philodendron scandens*). R. & N. TURNER

QUICK CHECK Climbing or trailing vines or large, leafy plants with heart-shaped leaves, sometimes deeply cut or lobed; entire plants contain irritant calcium oxalate crystals which cause intense burning of the mouth and throat if ingested; seldom life-threatening.

Fig. 199. Split-leaf Philodendron
(*Monstera deliciosa*). R. & N. TURNER

DESCRIPTION Climbing vines or large, leafy plants with woody stem bases, many growing up to 18 m (60 ft) or more high in the wild, clinging to the trunks of trees by means of aerial roots. When grown as potted indoor plants, they are usually much shorter, but can still attain a considerable length. The foliage and plant characteristics vary from one species to another, but in general the leaves are leathery and glossy or velvety, bright green to reddish, heart-shaped to arrowhead-shaped, and in some types, deeply lobed or cut. The leaf stalk forms a sheath around the stem. The plants seldom flower; when they do, male and female are produced separately on the same individual plant. The flower heads are Anthurium-like, with a leafy spathe subtending a long, slender spadix. One of the most commonly grown species is Heart-leaf Philodendron, or Sweetheart Plant (*P. scandens*). With thin, trailing stems and heart-shaped, pointed leaves up to 13 cm (5 in.) long, it is usually grown in hanging pots. Some Philodendron varieties produce red, burgundy, or cream-and-green variegated leaves. Another popular species is Split-leaf Philodendron, or Swiss-cheese Plant (*Monstera deliciosa;* syn. *Philodendron pertusum*), having large, interesting leaves that are deeply cut along the edges and perforated with irregularly placed holes.

OCCURRENCE Philodendrons are native to the tropical rain forests of Central and South America, and there are more than 200 species in the genera *Philodendron* and *Monstera*. Several of these are grown as potted ornamentals in North American homes and they are among the most popular of all house plants. In the southern United States they are also sometimes grown outdoors.

TOXICITY Philodendrons, like other members of the arum family, contain microscopic, sharp crystal bundles of calcium oxalate in their leaves, roots, stems, flowers, and unripe fruits. If swallowed, these crystals cause intense irritation and swelling of the mouth and throat, and can also cause skin and eye irritation on contact. The plants also contain unidentified toxic proteins. For a detailed description of symptoms of poisoning, see under Anthurium (p. 238).

TREATMENT See under Anthurium (p. 238); check for obstruction of air passage in cases of ingestion.

NOTES The trailing or climbing Heart-leaf Philodendron is a particular hazard to young children and pets because they can easily reach and chew on the hanging stems and leaves. Be especially watchful of babies as they learn to stand, crawl or use a walker. More than one mother has discovered her newly mobile baby with a Philodendron leaf in its mouth, pulled from a plant that only a week before had been out of reach.

The fleshy fruit of Split-leaf Philodendron (*Monstera deliciosa*) is edible, giving rise to other names for it, including Mexican Breadfruit and Fruit Salad Plant. The fruit takes a year or so to ripen, and must be fully mature before it can be eaten.

Poinsettia, or Christmas Flower
(Spurge Family)

Euphorbia pulcherrima
(Euphorbiaceae)

Fig. 200. Poinsettia (*Euphorbia pulcherrima*).
R. & N. TURNER

QUICK CHECK Low, bushy potted plant or large shrub with bright red, pink, or white colored floral bracts; entire plant exudes a milky juice when cut or bruised. Ingesting the plant may produce stomach upset; milky juice may irritate the skin.

DESCRIPTION Low and bushy plant with a woody base when grown as a potted house plant; when growing freely outdoors in warm areas or planted out in a greenhouse, Poinsettia is a large, woody shrub up to 3 m (10 ft) or more high. Indoors it is much smaller, and usually densely leafy. The leaves, 8–15 cm (3–6 in.) or more long, are alternate, bright green, oval or elliptical, smooth edged or shallowly lobed, and slightly hairy beneath. The familiar "flowers" of Christmas time are actually leafy, brightly colored bracts surrounding an open cluster of small yellow and red flowers, which mature into small capsules. The most common color of the flower bracts is bright vermilion-red, but there are also white, pink, and variegated forms. The entire plant exudes copious quantities of an acrid, milky juice when cut or bruised.

OCCURRENCE A native of Mexico and Central America, Poinsettia is used widely in North America as a potted Christmas-time ornamental. In warmer areas of the southern United States it is grown outdoors in containers or in gardens as a shrub or hedge plant.

TOXICITY The milky juice of Poinsettia, like that of other *Euphorbia* species, is an irritant, but it appears to be less potent than in other members of the genus. Much attention has centered on Poinsettia and its possible toxicity. Despite the many warnings against it, however, the only reported fatality from the plant was a single case in Hawaii in 1919. In recent years there have been hundreds of reported cases of children ingesting parts of Poinsettia. In most of these no symptoms or only minor ones were experienced, the most severe being vomiting, abdominal pain, and diarrhea. Chemical studies have shown the plant to be lacking the irritant diterpenes found in other Euphorbias. Handling Poinsettia, however, can be irritating to the skin.

TREATMENT Treat for symptoms; for skin irritation, wash thoroughly with warm, soapy water.

NOTES Poinsettia is a short-day flowering plant, and growing it for the North American Christmas market is a major industry. Modern plant breeding has developed shorter, bushier, longer-flowering hybrids, which will bloom year after year if allowed short days and long, dark nights in the fall.

Rosary Pea, or Jequirity Bean
(Bean, or Legume Family)

Abrus precatorius
(Fabaceae, or Leguminosae)

Fig. 201. Rosary Pea (*Abrus precatorius*). WALTER H. LEWIS

QUICK CHECK Climbing tropical vine whose attractive red and black seeds are often used as beads in novelty jewelry; seeds highly toxic if broken or chewed when ingested; may be fatal.

DESCRIPTION This weedy plant does not commonly grow in North America except in Florida. Its attractive but deadly seeds are the part

Fig. 202. Wreath of Rosary Pea seeds and fruits of Royal Poinciana (*Delonix regia*).
WALTER H. LEWIS

that usually find their way into North American households, in the form of beads on necklaces and novelty jewelry, inside rattles, and as good luck charms. The plant is a slender, twining vine growing 3 m (10 ft) or more tall, with pinnately compound leaves composed of many small leaflets, and red, lavender, or white pea-like flowers in small clusters. The fruiting pods are flat, broad, and hairy, about 4 cm (1.5 in.) long, each containing 4–8 shining, hard seeds about 6 mm (0.25 in.) long, which are half brilliant scarlet and half black.

OCCURRENCE Native to India, this plant has spread to most tropical and subtropical regions, and is a common weed in Florida, the Caribbean, Hawaii, and Guam. It is best known in North America for its attractive seeds, which are brought home by tourists as beads in necklaces and novelty jewelry.

TOXICITY The hard-coated mature seeds, if swallowed whole, usually pass through the digestive tract without harm. However, if chewed, broken, or immature, they release a highly toxic protein substance, a lectin or toxalbumin called abrin, which inhibits protein synthesis in growing cells of the intestinal wall and agglutinates red blood cells. Symptoms of poisoning occur after a latent period of many hours, or sometimes as much as three days, depending on the number of seeds swallowed, and the amount of chewing they underwent. Nausea, vomiting, diarrhea, severe abdominal pain, dilation of the pupils, and ulcerating and bleeding of the mouth and digestive tract are major symptoms. Loss of intestinal function and liver damage may occur, and in severe cases, convulsions, coma, and death. A single, well chewed seed can be fatal for a child, even with treatment. Seeds drilled to make beads are dangerous because the inner part is thus exposed.

TREATMENT Induce vomiting, or perform gastric lavage even if only a small amount has been eaten; follow with activated charcoal and a saline cathartic (repeat charcoal and cathartic every 6 hours for 24 hours); treat for symptoms; antacids may provide relief; maintain fluid and electrolyte balance; maintain good urine output, and alkalinize the urine; monitor blood pressure, heart, electrolytes, kidney and liver function; control convulsions with i.v. diazepam; if necessary institute parenteral alimentation; blood transfusions may be required.

NOTES The name Rosary Pea refers to the popular use of the seeds for rosaries. The specific name, *precatorius* is also derived from this use, since it originates from the Latin word *precator* meaning one who prays. The plant and its seeds have a wide variety of other local and colloquial names, most pertaining to the use of the seeds as beads. These include Crab's Eyes, Red Bead, Coral Bead, Prayer Beads, Love Bean, Indian Bead, and Seminole Bead.

References for Chapter V: Canadian Pharmaceutical Association 1984; Claus et al. 1970; Cooper and Johnson 1984; Der Marderosian and Roia 1979; Fuller and McClintock 1986; Hall 1973; Hardin and Arena 1974; Hedrick 1972; Hessayon 1985; Hill 1986; Kinghorn 1979; Kingsbury 1964; Kramer 1977; Lampe and McCann 1985; Leeuwenberg 1987; Lewis and Elvin-Lewis 1977; Mitchell and Rook 1979; Morton 1958, 1971.

APPENDIXES

APPENDIX 1

Common Fruits, Vegetables, and Beverage Plants Containing Potentially Harmful Toxins

Listed alphabetically by their common names; Almost all of these are harmless when used in normal quantities under normal circumstances and are toxic only when misused. References: Cooper and Johnson 1984; Fuller and McClintock 1986; Hardin and Arena 1974; National Academy of Sciences 1973.

Akee (*Blighia sapida*)—the covering around the unripe fruit, the fruit wall, and the seeds of this tropical fruit contain a compound called hypoglycin that can cause acute low blood sugar; sometimes fatal. In Jamaica, where this fruit commonly grows, Akee poisoning is called "vomiting sickness," and has been common in the past in times of food shortage. The ripe fruit is a favorite food.

Almond, bitter oil of (*Prunus dulcis*)—contains a cyanide-producing compound, amygdalin, most of which is removed in commercially produced oils (see under Apricot in text—p. 146).

Apple, seeds (*Malus* spp.)—contain a cyanide-producing compound; harmful if eaten in quantity (see under Apricot in Chapter IV, p. 146).

Apricot, kernels (*Prunus armeniaca*)—contain a cyanide-producing compound; harmful if eaten in quantity (see Chapter IV, p. 146).

Asparagus (*Asparagus officinalis*)—young shoots may cause skin irritation; berries not edible.

Avocado (*Persea americana*)—seeds, leaves, stems poisonous (see Chapter V, p. 223).

Banana and Plantain (*Musa* × *paradisiaca*)—fruit peel, and to a much lesser extent, the fruit pulp, contains amines (serotonin, tyramine, dopamine, norepinephrine) which cause an increase in blood pressure in animals.

Bean, Hyacinth (*Dolichos lablab*)—contains a cyanide-producing glycoside; must be thoroughly cooked (with 2–4 water changes).

259

Fig. 203. Asparagus (*Asparagus officinalis*).
R. & N. TURNER

Bean, Broad (*Vicia faba*)—eating fresh Broad Beans, cooked or uncooked, causes a blood disease, favism, in which red blood cells are destroyed and acute anaemia develops. It occurs only in susceptible people, particularly in those of the northern Mediterranean countries, who have an inherited deficiency of a particular enzyme (glucose–6-phosphate dehydrogenase). Broad Bean also contains a phytotoxin or lectin, and a glycoside, vicine.

Bean, Kidney (*Phasolus vulgaris*)—known to contain lectins.

Bean, Lima (*Phaseolus lunatus*)—some strains (not usually found in North America) contain dangerous quantities of a cyanide-producing glycoside.

Beet (greens) (*Beta vulgaris*)—high in soluble salts of oxalic acid and can cause calcium deficiency if eaten in high quantities in a calcium poor diet.

Brassicas (*Brassica* spp.)—the hot, pungent taste and odor of these and other mustards, and of mustard oils, is due to the presence of glucosinolate compounds, which are called goitrogens because they interfere with the uptake of iodine by the thyroid gland, and in extreme cases can lead to goiter. There is little evidence that these plants, when used as vegetables and condiments in moderation, cause problems for people with adequate iodine in their diets.

Broccoli (*Brassica oleracea*)—see Brassicas.

Brussels Sprouts (*Brassica oleracea*)—see Brassicas.

Buckwheat (*Fagopyrum esculentum*)—plant contains phototoxic phenolic compounds; can cause photosensitization in susceptible individuals.

Cabbage (*Brassica oleracea*)—see Brassicas.

Cashew (*Anacardium occidentale*)—raw nuts and kernels contain a cyanide-producing compound, but commercial processing removes this; oil may cause skin irritation in susceptible individuals.

Cassava (*Manihot esculenta*)—raw or improperly prepared tubers, and tuber peelings contain a cyanide-producing compound; potentially fatal.

Cauliflower (*Brassica oleracea*)—see Brassicas.

Celery (*Apium graveolens*)—plants, especially those contaminated with a mold (*Sclerotinia sclerotiorum*), produce phototoxins (psoralens, which are furanocoumarin compounds), sometimes causing serious dermatitis of workers handling celery.

Chamomile (*Anthemis nobilis*)—contains thujone, a volatile oil, which in concentrated form can cause delirium, hallucinations, and permanent brain damage; Chamomile is apparently safe in moderation, according to Tyler (1987).

Chard, Swiss (*Beta vulgaris*)—see Beet.

Cherries (*Prunus* spp.)—leaves, bark, and seed kernels contain a cyanide producing compound (see under Apricot in Chapter IV, p. 146).

Chili Pepper (*Capsicum annuum*)—see Chapter V (p. 244).

Chives (*Allium schoenoprasum*)—see Onion.

Cocoa, Chocolate (*Theobroma cacao*)—contains caffeine, a purine alkaloid, which has a stimulating effect on the central nervous system; see Coffee.

Coffee (*Coffea arabica*)—contains caffeine, a purine alkaloid, which has a stimulating effect on the central nervous system. Large quantities can lead to dizziness, pains, vomiting, rapid pulse, and lowered blood pressure. Chronic poisoning from continued high doses may result in headaches, restless, insomnia, tremors, and constipation.

Corn (*Zea mays*)—plants susceptible to nitrate accumulation and can cause poisoning in livestock; silage fermentation may produce lethal amounts of nitrogen dioxide and nitrogen tetraoxide gases,

Cress, garden (*Lepidium sativum*)—contains glucosinolates (see Brassicas).

Egg Plant (*Solanum melongena*)—plants, unripe fruits, and overripe fruits contain solanine and related compounds; see Potato in Chapter IV (p. 208).

Elderberry (*Sambucus* spp.)—see Appendix II.

Fig (*Ficus* spp.)—sap contains skin irritant; see Chapter V (p. 230).

Flax (*Linum* spp.)—seeds contain cyanide-producing glycosides, varying considerably in concentration depending on genetic strain, season, and climate; these are usually a problem only when linseed cake or meal is fed in quantity to livestock.

Garlic (*Allium sativum*)—see Onion.

Fig. 204. Horse Radish (*Armoracia rusticana*).
R. & N. TURNER

Horse Radish (*Armoracia rusticana*)—contains glucosinolates (see Brassicas).

Kale (*Brassica oleracea*)—see Brassicas.

Kiwi (*Actinidia chinensis*)—contains proteolytic enzymes; excessive use could cause irritation to mucous membranes.

Kohlrabi (*Brassica oleracea*)—see Brassicas.

Lima Bean—see Bean, Lima.

Mace (*Myristica fragrans*)—contains myristicin, a potent and dangerous hallucinogen, but harmful only in large amounts; see Nutmeg (Chapter V, p. 251).

Mango (*Mangifera indica*)—sap from stem at the base of the fruit contains a Poison Ivy-like toxin; carefully wash fruit, and do not eat the skin.

Mint, including Peppermint (*Mentha piperita*) and related species—contains menthol in its aromatic oil; in high doses menthol can be harmful to susceptible individuals.

Mustards (*Brassica juncea, B. nigra, Sinapis alba, S. arvensis*)—contain glucosinolates (see Brassicas).

Nutmeg (*Myristica fragrans*)—contains myristicin, a potent and dangerous hallucinogen, but harmful only in large amounts; see Chapter V (p. 251).

Onion (*Allium cepa* and related species)—Eaten in large amounts over a period of time, Onions can cause anemia, jaundice, and digestive disturbances in humans. They are also known to be harmful to cattle and horses. Onions, Chives, and Garlic contain several sulfur-containing volatile oils which cause irritation to the eyes and nose, and may also cause skin irritation.

Papaya (*Carica papaya*)—contains proteolytic enzymes which may cause irritation of mucous membranes with excessive use.

Parsnip (*Pastinaca sativa*)—contains phototoxic furanocoumarins; contact with foliage in sunlight can cause skin irritation, blistering, and discoloration.

Pea, field (*Pisum sativum*)—plant contains lectins, substances that agglutinate red blood cells and may, in large doses, interfere with the body's immune system.

Peach (*Prunus persica*)—bark, leaves, and seed kernels contain cyanide-producing compound (see under Apricot in Chapter IV, p. 146).

Peanut (*Arachis hypogaea*)—many people are allergic to peanuts; peanuts contaminated with certain molds (mainly *Aspergillus flavus*) contain potent carcinogens called aflatoxins.

Pepper, Black (*Piper nigrum*)—contains myristicin and safrole (see Nutmeg, and Sassafras in Appendix 2).

Peppermint (*Mentha piperita*)—see Mint.

Pepper, Chili—see Chili Pepper (*Capsicum annuum*) in Chapter V (p. 244).

Pineapple (*Ananas comosus*)—contains proteolytic enzymes which may cause irritation of mucous membranes with excessive use.

Plantain—see under Banana.

Potato (*Solanum tuberosum*)—greens, sprouts, and green tubers are toxic; see Chapter IV (p. 208).

Radish (*Rhaphanus sativus*)—contains glucosinolates (see Brassicas).

Rapeseed (*Brassica campestris, B. napus, B. carinata*)—seeds and oil contain erucic acid and glucosinolates (see Brassicas) (Note: canola oil is from Rapeseed varieties genetically developed in Canada, which are very low in erucic acid and glycosinolates and therefore safe to consume).

Rhubarb (*Rheum rhabarbarum*)—stalks contain oxalic acid; leaves very poisonous (see Chapter IV, p. 210).

Rutabaga (*Brassica napus*)—contains glucosinolates (see Brassicas).

Sage (*Salvia officinalis*)—contains thujone, which is toxic in high doses.

Soybean (*Glycine max*)—raw beans contain several enzyme inhibitors, proteins inhibiting the digestion of other proteins, and hemagglutinins, substances that agglutinate red blood cells; these are largely destroyed by cooking.

Spinach (*Spinacia oleracea*)—high in soluble salts of oxalic acid and can cause calcium deficiency if eaten in high quantities in a calcium poor diet.

Tea (*Camellia sinensis*)—contains caffeine, a purine alkaloid, which has a stimulating effect on the central nervous system.

Tomato (*Lycopersicon esculentum*)—plants contain a compound similar to solanine in green potatos (see under Potato in Chapter

IV, p. 208).

Turnip (*Brassica rapa*)—contains glucosinolates (see Brassicas).

Walnuts (*Juglans regia* and related species)—blackened hulls and moldy nuts have poisoned dogs eating them from the ground under the trees.

Watercress (*Nasturtium officinale*)—contains glucosinolates (see Brassicas).

**This book cannot replace the advice and assistance
of qualified medical personnel.
In all cases of suspected poisoning by plants, or any other substance,
immediate qualified medical advice and assistance should be sought.**

APPENDIX 2

Some Wild Edible Plants with Potentially Toxic Properties

Includes species that are no longer recommended as edible, but have been listed in edible plants books; some of these species are described in detail in Chapter III; References: Hardin and Arena 1974; Fuller and McClintock 1986.

Acorns (from Oaks—*Quercus* spp.)—contain tannins of the gallotannin class; for food use, acorns must be leached in running water to remove bitter tannins; leaves and shoots highly toxic to livestock (see Chapter III, p. 83).

Arrowgrass (*Triglochin maritima* and related species)—young vegetative leaf bases of *T. maritima* eaten by some Northwest Coast Indian groups (Turner 1975), but the plants under some conditions contain cyanide-producing glycosides and have caused illness and death in livestock.

Bay Laurel, California (*Umbellularia californica*)—leaves contain as much as 4 % dry weight as irritating oils, mainly umbellulone, a volatile oil. Inhaling this aromatic substance can cause headache and unconsciousness in some people.

Beechnut, European and American (*Fagus sylvatica* and *F. grandiflora*)—roasted nuts edible, but raw seed kernels contain a saponin-like substance and an alkaloid-like compound, fagin. If eaten in quantity (50 or more nuts), the kernels of the European Beech, and possibly American Beech as well, can produce headache, abdominal pain, vomiting, diarrhea, and extreme fatigue, symptoms which develop within an hour of eating the nuts and last up to 5 hours.

Fig. 205. American Beechnuts (*Fagus grandiflora*).
R. & N. TURNER

Bracken Fern (*Pteridium aquilinum*)—contains several toxic and carcinogenic substances; eating the young shoots or rhizomes of this species not recommended (see Chapter III, p. 69).

Broom (*Cytisus scoparius*)—flowers sometimes used in wine making, and roasted seeds as a coffee substitute, but the plant contains alkaloids and may cause digestive upset; use not recommended (see Chapter III, p. 86).

Buckeyes (*Aesculus* spp.)—some Native peoples used the thoroughly leached and roasted seeds as an emergency food, but the entire plants should be considered toxic (see Chapter III, p. 73).

Cherries, Wild (*Prunus* spp.)—bark, foliage, and seed kernels contain a potent cyanide-producing glycoside (see Chapter III, p. 74).

Chocolate Tips (*Lomatium dissectum*)—taproots and very young shoots eaten by some Native peoples in northwestern North America, but use not recommended; contain high concentrations of furanocoumarins, and used as fish poison.

Choke Cherry (*Prunus virginiana*)—see Cherries, Wild.

Clovers (*Trifolium* spp.)—Clover flowers and leaves are sometimes eaten or made into teas, and the rhizomes of some species are also eaten. However, Clovers should be used with caution and only in moderation, because they contain a number of toxic compounds, including infertility-causing estrogens, cyanide-producing glycosides, goitrogens, nitrates, and substances that can cause photosensitivity or coagulate the blood. They have poisoned grazing animals in many parts of the world, but mainly if eaten in large amounts or under unusual circumstances.

Cow Parsnip (*Heracleum lanatum*)—young leafstalks and budstalks peeled and eaten in spring by many Native peoples in North America (cf. Turner 1975, 1978; Kuhnlein and Turner 1987).

Fig. 206. Cow Parsnip (*Heracleum lanatum*).
R. & N. TURNER

However, plants contain several phototoxic furanocoumarins that can cause Poison Ivy-like dermatitis and discoloration of the skin if the plants are touched or handled followed by exposure to ultraviolet radiation from sunlight.

Dock (*Rumex* spp.)—leaves and stems used as potherb, but should be used only in moderation; contain soluble oxalates which can interfere with calcium uptake.

Elderberry (*Sambucus* spp.)—bark, twigs, leaves, and seeds contain cyanide-producing glycoside; ripe berries the only edible part (see Chapter III, p. 78).

Jack-in-the-pulpit (*Arisaema* spp.)—contains calcium oxalate crystals (see Chapter III, p. 115).

Kentucky Coffee Tree (*Gymnocladus dioica*)—contains alkaloid cyticine (see p. 79).

Labrador-tea (*Ledum palustre, L. glandulosum*)—contains andromedotoxins; apparently safe in weak tea solution, but should not be made too strong.

Fig. 207. Labrador-tea (*Ledum palustre*).
R. & N. TURNER

Lamb's-quarters (*Chenopodium album*)—high content of soluble oxalates, which can cause reduction of calcium when eaten in excess; may also cause phototoxicity when eaten in large quantities.

Lupines (*Lupinus* spp.)—roots of some species formerly eaten by Native peoples, but the plants contain lupine alkaloids; use not recommended (see Chapter IV, p. 201).

Mallow (*Malva* spp.)—greens and green fruits eaten, but plants known to cause livestock poisoning if eaten in quantity; toxin unknown (Fuller and McClintock 1986).

Marsh Marigold (*Caltha* spp.)—plants contain irritant protoanemonin; should never be eaten raw.

Mayapple (*Podophyllum peltatum*)—only fully ripe fruits edible (see Chapter III, p. 120).

Milkweed (*Asclepias* spp.)—young shoots and green fruits of *A. speciosa* and *A. syriaca* have been eaten as potherbs, but all species should be treated with caution (see Chapter III, p. 123).

Mountain-Ash (*Sorbus* spp.)—fruits used to make jellies and wine, but seeds should be strained out; contain a cyanide-producing glycoside.

Mulberry, Red (*Morus rubra*)—unripe fruits and milky sap in leaves and stems cause hallucinations and digestive upset.

Onions, Wild (*Allium* spp.)—use in moderation; some species can cause digestive upset and some species have caused breakdown of red blood cells in livestock.

Orache (*Atriplex* spp.)—may be phototoxic if eaten in excessive amounts.

Pawpaw (*Asimina triloba*)—fruits may cause skin irritation.

Pigweed (*Amaranthus* spp.)—plants may accumulate oxalates and nitrates; known to cause fluid accumulation around the kidneys of livestock eating the plants.

Pokeweed (*Phytolacca americana*)—roots and raw leaves and fruits have caused serious, fatal poisoning. Fully cooked greens and berries have been eaten in the past, but use not recommended because of presence of cell-destroying mitogens (see Chapter III, p. 131).

Purslane (*Portulaca oleracea*)—plants may accumulate oxalates and nitrates; use only in moderation.

Saskatoon berry (*Amelanchier alnifolia* and related species)—fruits safe and edible, but leaves, bark, and seeds contain cyanide-producing glycosides and can be toxic (seeds present in fruits are not harmful with normal use).

Sassafras (*Sassafras albidum*)—root bark used as tea; its volatile oil contains safrole, which is known to cause tumors in rats; hence the use of safrole for flavoring has been banned by the United States Food and Drug Administration.

Sedum (*Sedum* spp.)—various species contain oxalic acid and soluble oxalates; use only in moderation.

Serviceberry—see under Saskatoon.

Skunk-Cabbage, Western (*Lysichitum americanum*)—entire plant contains needle-like calcium oxalate crystals; rhizomes eaten by some West Coast Native peoples, but must be prepared specially and cooked for a long time.

Soapberry (*Shepherdia canadensis*)—berries contain bitter, sudsing saponins that allow them to be whipped up with water into a favorite confection of Native peoples of British Columbia (Turner 1975); no reports of poisoning from their use, but should be used only in moderation.

Sourgrass, or Sheep Sorrel (*Rumex acetosella*)—plants contain oxalic acid and soluble oxalates; use only in moderation.

Sweet Flag (*Acorus calamus*)—volatile oil contains asarone, which causes illness and tumor growth in rats; calamus oil has now been withdrawn from use as a flavoring ingredient in many countries.

Walnut, Black (*Juglans nigra*)—hulls and moldy nuts are toxic to livestock.

Water Parsnip (*Sium suave*)—Native peoples consider the parsnip-like roots edible (Turner 1978), but accidents have occurred when they have been confused with the highly toxic Water Hemlock (*Cicuta* spp.) (see Chapter III, p. 141).

Wild Ginger (*Asarum* spp.)—volatile oil contains asarone, which can cause illness and tumor growth in rats in high concentrations.

Wood-Sorrel (*Oxalis* spp.)—plants contain oxalic acid and soluble oxalates; use only in moderation.

Yaupon tea (*Ilex vomitoria*)—causes vomiting at high concentrations; see under Holly (p. 155).

**This book cannot replace the advice and assistance
of qualified medical personnel.
In all cases of suspected poisoning by plants, or any other substance,
immediate qualified medical advice and assistance should be sought.**

APPENDIX 3

Some Common Plants Causing Skin Irritation and/or Mechanical Injury

There are several types of skin irritation, or dermatitis caused by plants. Some are mechanical, caused by injury from thorns, barbs or spines. Others, such as those caused by Buttercups, Spurges, and Stinging Nettles, are from irritant chemicals present in the juice, sap, or hairs. In other cases, an allergic reaction results from a person being sensitized to a particular plant. Some plants, such as Poison Ivy and its relatives, are allergy-causing in most individuals after an initial exposure. Some plants are phototoxic. These cause the skin to be hypersensitive to sunlight, and skin irritation occurs after exposure to these plants with subsequent exposure to sunlight (see discussion in Chapter I). References and complete listings: Fuller and McClintock (1986); Hardin and Arena (1974); Lampe and Fagerström (1968); Mitchell and Rook (1979); see also Chi-Kit et al. (1980). Many of the plants listed also cause eye irritation or injury if allowed to come in contact with the eyes.

Allamanda (*Allamanda cathartica*)—all parts
Anemone (*Anemone* spp., *Pulsatilla* spp.)—leaves, flowers
Arums (many members of arum family: Araceae)—leaves, stems, roots
Asparagus (*Asparagus officinalis*)—young shoots
Barberry (*Berberis* spp.)—spines
Bird Pepper (*Capsicum frutescens*)—fruits
Bleedingheart and related species (*Dicentra* spp.)—all parts
Bloodroot (*Sanguinaria canadensis*)—sap
Blue Cohosh (*Caulophyllum thalictroides*)—roots
Borage (*Borago* spp.)—prickly hairs
Boxwood (*Buxus sempervirens*)—leaves
Burdock (*Arctium* spp.)—burs
Buttercups (*Ranunculus* spp.)—leaves, flowers
Cactus (*Opuntia* spp. and others)—barbed spines and bristles
Cardinal Flower (*Lobelia cardinalis*)—sap, leaves
Carrot, Wild (*Daucus carota*)—leaves phototoxic
Cashew (*Anacardium occidentale*)—oil, nutshells; strongly allergenic
Castor Bean (*Ricinus communis*)—plant
Cedar, Red- and Yellow- (*Thuja plicata, Chamaecyparis nootkatensis*)—boughs, wood allergenic
Celandine (*Chelidonium majus*)—red sap
Celery (*Apium graveolens*)—phototoxic (especially when contaminated with a mold, *Sclerotinia sclerotiorum*)

Fig. 208. Burdock (*Arctium minus*).
R. & N. TURNER

Fig. 209. Prickly-Pear Cactus (*Opuntia fragilis*).
R. & N. TURNER

Fig. 210. Devil's-Club
(*Oplopanax horridus*).
R. & N. TURNER

Century Plant (*Agave* spp.)—sap
Chili Pepper (*Capsicum annuum*)—fruits
Chrysanthemum (*Chrysanthemum* spp.)—plants
Citrus fruits (*Citrus* spp.)—peel, thorns
Clematis (*Clematis* spp.)—leaves
Cocklebur (*Xanthium strumarium*)—plants
Cohosh—see Blue Cohosh
Comfrey (*Symphytum* spp.)—hairs on leaves, stems
Cow Parsnip (*Heracleum lanatum*)—sap of leaves, stems phototoxic;
    potentially serious
Croton (*Codiaeum* spp.)—plant
Devil's Walkingstick, or Hercules' Club (*Aralia spinosa*)—bark
Devil's Club (*Oplopanax horridus*)—spiny stems can cause serious
    allergy
Dittany, or Gas Plant (*Dictamnus albus*)—plants phototoxic
Dock (*Rumex* spp.)—plants phototoxic
Feverfew (*Parthenium hysterophorus*)— commonly allergenic with
    skin contact

Figs (*Ficus* spp.)—sap
Giant Hogweed (*Heracleum mantegazzianum*)—sap of leaves, stems phototoxic; potentially serious
Ginkgo (*Ginkgo biloba*)—seeds
Gooseberry (*Ribes* spp.)—spines
Grasses [many species, including Cheat Grass, or Downy Brome (*Bromus tectorum*); Needle Grass (*Stipa* spp.); Three-awned Grass (*Aristida longiseta*); Wild, or Foxtail Barley (*Hordeum jubatum*); and Wild Oats (*Avena fatua*)]—Awned grass fruits and bracts can become embedded in the fur, ears, eyes, and nostrils of pets and livestock; sometimes inadvertently swallowed; can cause choking
Hawthorn (*Crataegus oxyacantha* and related species)—thorns
Heliotrope (*Heliotropium* spp.)—leaves
Holly (*Ilex* spp.)—prickly leaves
Hops (*Humulus lupulus*)—juice allergenic
Hyacinth (*Hyacinthus orientalis*)—plants, bulbs
Indian Hellebore (*Veratrum viride*)—leaves
Indian Tobacco (*Lobelia inflata*)—juice, leaves
Iris (*Iris* spp.)—juice
Ivy, English (*Hedera helix*)—leaves
Jack-in-the-Pulpit (*Arisaema* spp.)—leaves, roots
Jessamine, Yellow (*Gelsemium sempervirens*)—leaves, stems
Jimsonweed (*Datura* spp.)—leaves, flowers
Junipers (*Juniperus* spp.)—boughs, wood allergenic
Knotweeds (*Polygonum* spp.)—leaves phototoxic
Ladyslippers (*Calceolus* spp., *Cypripedium* spp.)—leaves
Lily-of-the-Valley (*Convallaria majalis*)—leaves, roots
Manchineel (*Hippomane mancinella*)—milky juice; potentially serious
Mango (*Mangifera indica*)—sap from fruit stem allergenic
Marigold (*Tagetes* spp.)—plants phototoxic
Marsh-Elder, or Copperweed (*Iva* spp.)—plants phototoxic
Mayapple (*Podophyllum peltatum*)—roots
Mesquite (*Prosopis* spp.)—spines
Mulberry (*Morus rubra*)—leaves, stem
Oleander (*Nerium oleander*)—leaves
Osage Orange (*Maclura pomifera*)—milky juice
Nettle, Stinging—see Stinging Nettle
Nettle, Wood—see Wood Nettle
Palm, Date (*Phoenix* spp.)—leaf stalk thorns , sharp-tipped leaves
Papaya (*Carica papaya*)—sap
Parsnip (*Pastinaca sativa*)—leaves phototoxic
Pawpaw (*Asimina triloba*)—fruits
Plum, Sloe, or Blackthorn (*Prunus spinosa*)—spines
Plumeria, or Frangipani (*Plumeria rubra*)—sap
Poinsettia (*Euphorbia pulcherrima*)—sap

Poison Hemlock (*Conium maculatum*)—leaves

Poison Ivy (*Toxicodendron radicans*)—severely allergenic to many people (see p. 87).

Poison Oak (*Toxicodendron quercifolium* and *T. diversilobum*)—severely allergenic to many people (see p. 89).

Poison Sumac (*Toxicodendron vernix*)—severely allergenic to many people (see p. 85).

Poisonwood (*Metopium toxiferum*)—all parts; severely allergenic

Primrose (*Primula* spp.)—commonly allergenic

Privet (*Ligustrum vulgare*)—leaves

Puncture Vine (*Tribulus terrestris*)—spiny fruits

Ragwort (*Senecio* spp.)—leaves phototoxic

Rose (*Rosa* spp.)—spines

Russian Thistle (*Salsola kali*)—spines

Scarlet Pimpernel (*Anagallis arvensis*)—leaves

Spurges (*Euphorbia* spp.)—milky juice, spines of some species; potentially serious

Spurge Nettle (*Cnidoscolus stimulosus;* syn. *Jatropha stimulosa*)—stinging hairs; very painful, potentially serious

Stinging Nettles (*Urtica* spp.)—stinging hairs; very painful

Sweet Cicely, Wild (*Osmorhiza* spp.)—spine-like fruits can become embedded in skin or ears of pets; if swallowed can cause choking

Tansy (*Tanacetum vulgare*)—plants phototoxic

Tomato (*Lycopersicon esculentum*)—plants

Trumpet Creeper (*Campsis radicans*)—leaves, flowers; potentially serious

Tulip (*Tulipa* spp.)—leaves, bulbs

Wood Nettle (*Laportea canadensis*)—stinging hairs; very painful

Yarrow (*Achillea millefolium*)—plants phototoxic

Yucca (*Yucca* spp.)—sharp, pointed leaves

Fig. 211. Stinging nettle (*Urtica dioica*). R. & N. TURNER

APPENDIX 4

Honey Poisons

Honey is flower nectar that has been gathered, modified, concentrated, and stored by honeybees. Poisoning by honey is rare, either because bees often avoid flowers of poisonous plants, or because the bees themselves are poisoned directly before honey can be produced. Commercial honey producers are careful to ensure that poisonous honey does not reach the market; the main danger is in eating home-produced honey where the bees' nectar sources have not been monitored. Flowers of the following plants found in North America are known to yield poisonous honey. A thorough discussion is given by Patwardhan and White (1973).

>Azaleas (*Rhododendron* spp.)
>Bog Rosemary (*Andromeda* spp.)
>Buckeye (*Aesculus* spp.)
>Buttercups (*Ranunculus* spp.)
>   and related species; various
>   members of this family,
>   according to Fuller and
>   McClintock (1986)
>Henbane (*Hyoscyamus niger*)
>Jessamine, Yellow (*Gelsemium*
>   *sempervirens*)
>Jimsonweed (*Datura* spp.)
>Laurels, Mountain and Swamp
>   (*Kalmia* spp.)
>Milkvetches (*Astragalus* spp.)
>Oleander (*Nerium oleander*)
>Pieris (*Pieris japonica*)
>Ragwort (*Senecio jacobea*)
>Rhododendrons (*Rhododendron*
>   spp.)

APPENDIX 5

Milk Poisons

The following toxic plants are known to taint milk of browsing or grazing dairy cows and/or goats. In addition to these species, many other toxic plants can reduce milk yield. In only a few cases (e.g., in White Snakeroot), has a definite link been established between tainted milk and poisoning in people drinking it. References: Cooper and Johnson 1984; Hardin and Arena 1974.

Autumn Crocus (*Colchicum autumnale*)
Bracken Fern (*Pteridium aquilinum*)
Buttercups (*Ranunculus* spp.)
Fool's Parsley (*Aethusa cynapium*)
Garlics (*Allium* spp.)
Hellebore (*Helleborus* spp.)
Henbane (*Hyscyamus niger*)
Horsetail (*Equisetum* spp.)
Ivy, English (*Hedera helix*)
Laburnum (*Laburnum anagyroides*)
Lupines (*Lupinus* spp.)
Oaks (*Quercus* spp.)
Poison Hemlock (*Conium maculatum*)
Potato, greens, sprouts, green tubers (*Solanum tuberosum*)
Ragwort (*Senecio jacobea* and related species)
Snakeroot, White (*Eupatorium rugosum*)
Wood-Sorrel (*Oxalis* spp.)
Yew (*Taxus* spp.)

APPENDIX 6

Some Medicinal Herbs of Questionable Safety

The plants are listed in alphabetical order of their common names. Information on safety is derived mainly from Duke (1985) and Tyler (1987). The reader is referred to these authoritative books on medicinal herbs for details of these and other potentially harmful herbs. The comments generally pertain to internal consumption of the herbs, or solutions containing them. Although many of these herbs may be used without harm when prescribed by properly trained and experienced herbalists, none should be taken without extreme care and awareness of their potential dangers. Those suggested to be "very dangerous" should not be used at all in most circumstances.

Akee (*Blighia sapida*)—plant considered very dangerous for herbal use.

Angelica (*Angelica archangelica*)—Tyler (1987) considers this plant safe when used as a flavoring, but unsafe if used medicinally in more concentrated form.

Apricot pits, or Laetrile (*Prunus armeniaca*)—Tyler (1987) considers it dangerous and ineffective as an anticancer agent; Duke (1985) also rates it as potentially dangerous (see p. 146 of text).

Arnica, or Mountain Tobacco (*Arnica montana*)—plant considered generally unsafe for herbal use.

Autumn Crocus (*Colchicum autumnale*)—entire plant considered very dangerous for herbal use (see p. 181 of text).

Belladonna (*Atropa belladonna*)—plant considered very dangerous for herbal use (see p. 183 of text).

Broom (*Cytisus scoparius*)—plant potentially unsafe for herbal use (see p. 86 of text).

Canaegre (*Rumex hymenosepalus*)—due to high tannin content, Tyler (1987) considers root potentially carcinogenic in long-term herbal use; Duke (1985) also rates it as potentially unsafe.

Castor Bean (*Ricinus communis*)—entire plant, especially seeds, considered very dangerous for herbal use (see p. 224 of text).

Chaparral, or Creosote Bush (*Larrea tridentata*)—Tyler (1987) considers leaves and twigs unsafe for use as an alterative and anticancer agent. Rated relatively safe by Duke (1985).

Chinaberry (*Melia azedarach*)—plant very dangerous for human use (see p. 154 of text).

Cohosh, Black (*Cimicifuga racemosa*)—plant potentially unsafe for herbal use .

Fig. 212. Blue Cohosh (*Caulophyllum thalictroides*).
WASHINGTON UNIVERSITY, ST. LOUIS

Fig. 213. Blue Cohosh, in fruit. MARY W. FERGUSON

Fig. 214. Comfrey (*Symphytum officinale*).
R. & N. TURNER

Cohosh, Blue (*Caulophyllum thalictroides*)—plant potentially unsafe for herbal use.

Coltsfoot (*Tussilago farfara*)—Tyler (1987) considers leaves and flower heads unsafe for use as a demulcent and against coughs; may contain liver-damaging pyrrolizidine alkaloids (Bergner 1989). Rated relatively safe by Duke (1985).

Comfrey (*Symphytum officinale*)—Tyler (1987) considers plant potentially unsafe as a general healing agent when taken internally; may contain pyrrolizidine alkaloids, which can cause permanent and irreversible liver damage (Bergner 1989).

Cycad, or Sago Cycas (*Cycas revoluta*)—plant considered very dangerous for herbal use.

Daffodil (*Narcissus tazetta*)—plant considered very dangerous for herbal use (see p. 188 of text).

Darnel (*Lolium temulentum*)—entire plant considered very dangerous for herbal use (see p. 108 of text).

Daphne (*Daphne mezereum*)—plant and berries considered very dangerous for herbal use (see p. 166 of text).

Datura (*Datura* spp.)—plants considered very dangerous for herbal use (see p. 118 of text).

Dong Quai (*Angelica polymorpha*)—Duke (1985) and Tyler (1987) consider root potentially unsafe for herbal use.

Eyebright (*Euphrasia officinalis*)—Tyler (1987) considers plant unsafe and ineffective for treatment of eye diseases. Duke (1985) rates it as relatively safe.

Fool's Parsley (*Aethusa cynapium*)—plant considered very dangerous for herbal use.

Foxglove (*Digitalis purpurea*)—entire plant considered very dangerous for herbal use (see p. 192 of text).

Glory Lily (*Gloriosa superba*)—entire plant considered very dangerous for herbal use (see p. 194 of text).

Gotu Kola (*Centella asiatica*)—safety and efficacy of herbal use questionable.

Heliotrope (*Heliotropium* spp.)—dangerous for internal use; contains pyrrolizidine alkaloids, which can cause permanent and irreversible liver damage (cf. Fuller and McClintock 1986; Bergner 1989).

Hellebore, or Christmas Rose (*Helleborus niger*)—entire plant very dangerous for herbal use (see p. 195 of text).

Henbane (*Hyoscyamus niger*)—entire plant considered very dangerous for herbal use (see p. 197 of text).

Hound's Tongue (*Cynoglossum officinale*)—dangerous for internal use; contains pyrrolizidine alkaloids, which can cause permanent and irreversible liver damage (cf. Fuller and McClintock 1986).

Hydrangea (*Hydrangea* spp.)—safety and efficacy of herbal use questionable according to Tyler (1987), but Duke (1985) considers it relatively safe (see p. 168 of text).

Indian Hellebore (*Veratrum viride*)—entire plant considered very dangerous for herbal use (see p. 110 of text).

Indian Tobacco, or Lobelia (*Lobelia inflata*)—entire plant potentially dangerous for herbal use (see p. 114 of text).

Jessamine, Yellow (*Gelsemium sempervirens*)—entire plant considered very dangerous for herbal use (see p. 176 of text).

Jimsonweed (*Datura stramonium*)—entire plant considered very dangerous for herbal use (see p. 118 of text).

Juniper (*Juniperus* spp.)—safety of berries for use as diuretic questionable .

Laburnum (*Laburnum anagyroides*)—entire plant considered very dangerous for herbal use (see p. 158 of text).

Laurel, Mountain (*Kalmia latifolia*)—entire plant considered very dangerous for herbal use (see p. 80 of text).

Licorice (*Glycyrrhiza* spp.)—herbal use of rhizomes and roots potentially unsafe.

Life Root (*Senecio aureus*)—entire plant potentially unsafe for herbal

Fig. 215. Juniper (*Juniperus scopulorum*). R. & N. TURNER

use (see p. 133 of text); contains pyrrolizidine alkaloids (Bergner 1989).

Mandrake (*Mandragora officinarum*)—plant considered very dangerous for herbal use.

Mercury, Annual (*Mercurialis annua*)—plant considered very dangerous for herbal use.

Mistletoe (*Phoradendron* spp., *Viscum* spp.)—leaves considered unsafe for herbal use (see p. 250 of text).

Monkshood (*Aconitum* spp.)—plants considered very dangerous for herbal use (see p. 204 of text).

Mormon Tea (*Ephedra nevadensis*)—long-term use of stems as beverage, or herbal medicine potentially unsafe.

Nux-vomica, or Strychnine (*Strychnos nux-vomica*)—plant considered very dangerous for herbal use.

Oleander (*Nerium oleander*)—entire plant considered very dangerous for herbal use (see p. 233 of text).

Pennyroyal, European (*Mentha pulegium*)—herbal use of leaves or oil considered unsafe.

Peony (*Paeonia officinalis*)—entire plant considered very dangerous for herbal use.

Periwinkle (*Vinca minor*)—entire plant considered very dangerous for herbal use.

Physic Nut (*Jatropha curcas*)—entire plant considered very dangerous for herbal use.

Poinsettia (*Euphorbia pulcherrima*)—plant considered very dangerous for herbal use (see p. 254 of text).

Poison Hemlock (*Conium maculatum*)—entire plant considered very dangerous for herbal use (see p. 128 of text).

Pokeweed (*Phytolacca americana*)—entire plant considered very dangerous for herbal use (see p. 131 of text).

Ragwort (*Senecio* spp.)—see Life Root.

Rosary Pea (*Abrus precatorius*)—seeds and entire plant very dangerous for herbal use (see p. 255).

Sandbox Tree (*Hura crepitans*)—entire plant considered very dangerous for herbal use (see p. 161 of text).

Sage (*Salvia officinalis*)—leaves considered unsafe when used in concentrated doses in herbal medicine, but safe when used in moderation as flavoring (Tyler 1987). Duke (1985) considers it a safe herb.

Sagebrush (*Artemisia* spp.)—see Wormwood.

Sassafras (*Sassafras albidum*)—root bark considered unsafe for medicinal or beverage use, according to Tyler (1987), but rated as relatively safe by Duke (1985).

Spurges (*Euphorbia* spp.)—plants considered very dangerous for herbal use (see p. 213 of text).

Squill, Red or Sea-Onion (*Drimia maritima;* syn. *Urginea maritima*)—entire plant considered very dangerous for herbal use.

Sweet Flag, or Calamus (*Acorus calamus*)—rhizome considered potentially unsafe for herbal use.

Tansy (*Tanacetum vulgare;* syn. *Chrysanthemum vulgare*)—leaves and tops possibly unsafe for herbal use according to Tyler (1987), but rated relatively safe by Duke (1985).

Thornapple (*Datura innoxia*)—entire plant considered very dangerous for herbal use (see p. 118 of text).

Tobacco (*Nicotiana* spp.)—entire plants considered very dangerous for herbal use (see p. 218 of text).

Water Hemlock (*Cicuta* spp.)—entire plants considered very dangerous for herbal use (see p. 141 in text).

Wormwood, or Absinthe (*Artemisia absinthium*)—leaves and tops considered unsafe for use as a deworming agent or a tonic, for mind altering action, or as flavor, according to Tyler (1987).

Fig. 216. Sweet flag (*Acorus calamus*). WALTER H. LEWIS

GLOSSARY

Achene – small, dry indehiscent one-seeded fruit with tight, thin outer wall.

Algal bloom – rapid growth of algae in water, resulting in visible coating or coloring of the water.

Alternate – leaves or buds arising first on one side of the stem and then on the other (compare Opposite).

Annual – plant growing from seed, flowering, producing seeds, and dying all in a single season.

Anther – male, or pollen-bearing structure of a flower, usually borne on a stalk, or filament (see also Stamen).

Apical – at the tip of a branch or stem.

Arborescent – of tree-like habit.

Aril – fleshy covering around a seed.

Axil – angle formed between the stem axis and attached leaf or other appendage.

Berry – simple, fleshy, usually indehiscent (not splitting open) fruit with one or more seeds (e.g., Gooseberry, Tomato).

Biennial – plant growing from seed and living for two years, with flowering and seed production usually occurring in the second year (e.g., Foxglove).

Bisexual – having both sexual reproductive structures (male and female) produced on the same plant.

Bloom – a flower (see also Algal bloom); also a whitish, waxy covering on fruits and other plant structures.

Bract – modified leaf; some bracts growing near a flower may be more showy than the flower itself (e.g., Poinsettia, where the red "flowers" are actually bracts).

Bulb – modified short, underground stem surrounded by a row of fleshy leaves with thickened bases (e.g., Onion).

Calyx – sterile outer whorl of flower parts composed of sepals, usually covering an unopened flower bud.

Capsule – dry, dehiscent (splitting open) fruit in the form of a case containing seeds.

Carpel – one of the units comprising a pistil or ovary (female part) in a flower; a simple pistil has one carpel, a compound pistil two or more united carpels.

281

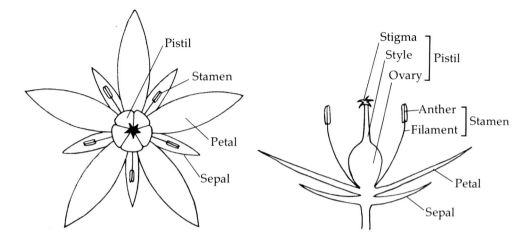

Fig. 217. Parts of a typical, regular, 5-parted flower. Left—top view; right—side view.

Catkin – spike-like inflorescence (flower cluster), usually containing scaly bracts.

Chaff – small, thin, dry, membranous scale or bracts, such as around the grains of grasses.

Clasping – partially or completely surrounding a stem, as the bases of some leaves.

Compound leaf – two or more leaflets attached to the leaf stalk (e.g., Walnut, Horse Chestnut, or Lupine).

Convex (cap of mushrooms) – regularly rounded; broadly obtuse.

Corm – swollen, usually spherical or rounded underground stem, capable of producing a new plant.

Corolla – the ring of petals in a flower, usually the most conspicuous part of the flower.

Crown – the region of a plant where the root and shoot join; the branching part of a tree.

Crustose – lichen growth form in which the thallus (main structure of lichen) adheres to the substrate on which it grows.

Cup – in mushrooms, a cup-like remnant of the universal veil found at the stem base (also called Volva).

Cylindric – a shape having the same diameter throughout; tube-like.

Dehiscent – opening by valves or along regular lines, as in ripe seed capsules or some anthers.

Drupe – type of fleshy fruit generally containing one seed (e.g., Cherry).

Entire leaf – leaf whose margins are smooth (compare Serrated).

Evergreen – having foliage that remains green and functional through more than one growing season.

Fibrillose (stringy) – composed of parallel fibers, especially of the cap in some species of mushrooms.

Fibrils – minute hairs.

Follicle – dry fruit of one carpel that splits along one suture line (e.g., Columbine fruit).

Frond – leaf of a fern or palm, or similar leaf-like structure.

Fruticose – lichen growth form in which the thallus (main structure of lichen) is generally erect and branching.

Genus (pl. genera) – A group of closely related plants containing one or more species.

Gills – radiating plate-like spore-bearing structures found on the undersurface of the cap of some mushrooms.

Glabrous – smooth, not hairy.

Herb – vascular plant with little, if any secondary growth; not woody.

Herbaceous – having the characteristics of an herb (not woody).

Hybrid – plant with parents which are genetically distinct. The parent plants may be different cultivars, varieties, species, or genera; the hybrid usually exhibits a mixture of traits of its parents.

Hyphae (singular, hypha) – minute tubular structures forming the mycelium and fruiting bodies of fungi.

Indehiscent – not opening, or not opening by valves or along regular lines, as some fruits and anthers.

Inflorescence – term for cluster or grouping of flowers on an axis or stalk; also, the mode of arrangement of the flower on a plant.

Involute – mushroom cap with inrolled margin.

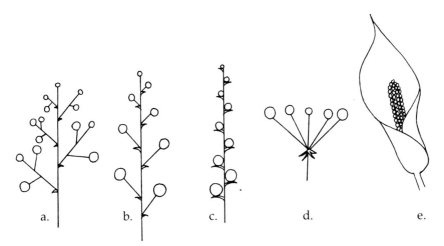

Fig. 218. Some common types of flower arrangements. a. panicle (loose cluster); b. raceme (elongated cluster with stalked flowers); c. spike (elongated cluster with stalkless flowers); d. umbel (umbrella-like cluster); e. arum type, with spathe surrounding club-like spadix.

Keel – the two united, lowermost petals of flowers belonging to the bean, or legume family.

Latex – sap produced or exuded by plants.

Ligulate – strap-shaped, or tongue-shaped.

Ligulate flower (see **Ray flower**).

Luminescence – emission of light from a living organism.

Mycelium – mass of hyphae making up the main growing body of a fungus; typically embedded in soil, wood, etc.

Node – place on the stem where one or more leaves or branches are attached.

Nut – hard, (usually) one-seeded fruit.

Nutlet – a small or diminutive nut, similar to an achene but with a harder, thicker wall.

Opposite – leaves or buds which are borne in pairs along the stem (compare Alternate). ·

Palmate leaf – five or more lobes arising from one point in a hand-like arrangement (e.g., typical maple leaf).

Parasite – organism that derives its nutrients and energy from a living host.

Pedicel – stalk bearing reproductive structures (flower stem).

Pistil – female flowering structure, typically consisting of an ovary (containing one or more eggs, later to be fertilized and grow into seeds), a neck-like style, and a stigma which receives pollen from the male flowering structure, or anther. Pistils and anthers may be contained within the same flower or in separate flowers.

Perennial – plant living from year to year for three or more years under normal conditions.

Prostrate – lying flat on the ground.

Raceme – unbranched, elongated inflorescence (flower cluster) with stalked flowers, the earliest blooming at the bottom, the least mature at the top.

Ray flower – ligulate flower, with corolla flattened and strap-like above a very short tube (e.g., the outer "petal" of a daisy, actually comprising a single flower in a composite head).

Receptacle – part of the floral axis supporting floral parts.

Red Tide or **Red Water** – concentration of toxic species of algae that cause a reddish coloration in water.

Rhizome – creeping, often fleshy underground stem from which new plants may arise.

Ring (Annulus) – in mushrooms, a membranous ring found on mature stems of some species; remnant of partial veil.

Runner – creeping above-ground stem which may produce small plantlets along its length; also called a stolon.

Sclerotium (pl. **sclerotia**) – resting body of some fungi, composed of a hardened mass of hyphae usually rounded in shape.

Serrate leaf – toothed around the margin.

Sessile – lacking a stem or stalk.

Simple leaf – one leaf blade attached to a leaf stalk (compare Compound leaf).

Spadix – fleshy flower spike bearing many tiny florets; the type of flower spike found in members of the arum family, usually enclosed or subtended by a leafy sheath, or spathe.

Fig. 219. Common leaf types; simple leaves. Top left—palmately lobed (English Ivy); top right—pinnately lobed (Oak); middle left—toothed or serrated margin (Cherry); middle center—spiny margin (Holly); middle right—smooth or entire margin, oval shaped (Waxberry); lower left—heart shaped, or cordate (Morning Glory); lower center—ellipse-shaped (Privet); lower right—inversely lance-shaped, or oblanceolate (Daphne Laurel).

Fig. 220. Common leaf types; compound leaves. Top left—deeply lobed, palmately compound (Monkshood); top right—finely divided (three times pinnately compound). (Poison Hemlock); top center—palmately compound, smooth margined leaflets (Lupine); center left—palmately compound, toothed margined leaflets (Horse Chestnut); center right—pinnately compound, coarsely toothed leaflets (Clematis); lower left—three-parted (Laburnum); lower right—pinnately compound, with terminal tendril (Vetch).

Spathe – large, often colored or leafy bract surrounding the spadix in members of the arum family.

Species – plants which are genetically similar and breed true from seed.

Spicate – spike-like.

Spike – unbranched, elongated inflorescence (flower cluster) in which the flowers are sessile (stalkless).

Spikelet – a secondary spike.

Stamen – male reproductive organ of a flower, consisting of a pollen-bearing structure (anther) borne on a stalk, or filament.

Stem, or Stipe – stalk-like portion of a mushroom's fruiting body; main stalk of any plant.

Stipule – one of a pair of appendages at the base of a petiole, or leaf stalk, of some plants.

Stolon (see **Runner**).

Striate – having lines or minute grooves; common in mushrooms on the upper surface of the cap near its margin, or on the stem.

Succulent – plants having thick, fleshy leaves and/or stems.

Taproot – strong, sometimes swollen or fleshy root which grows vertically into the soil.

Tendril – thread-like stem or leaf which clings to any nearby support (e.g., pea vines)

Thallus – having a simple plant body without differentiation into leaves or leaf-like structures.

Thallose – in lichens, having a simple, flattened thallus.

Tomentose – densely matted with a covering of soft hairs.

Tuber – swollen underground stem, such as produced by Dahlias and Potatos, which can sprout and produce new plants.

Umbel – umbrella-shaped inflorescence (flower cluster) in which pedicels (stalks of flowers or flower clusters) radiate from a common point, like the ribs of an umbrella.

Umbo – a raised, conical mound on the center of the cap of some mushrooms.

Undulate – wavy.

Unisexual – having only one type of sexual structure (either male or female) produced by any one individual.

Universal veil – tissue surrounding a developing mushroom button.

Variegated leaf – green leaf which is blotched, edged, or spotted with yellow, white, cream, or red.

Viscid (sticky) – (pertaining to mushroom cap) slimy or slippery surface when wet, becoming sticky as it starts to dry.

Volva – see Cup.

Whorl – leaves, petals, or branches arranged in a ring manner.

REFERENCES

Ainsworth, G. D. 1952. *Medical Mycology.* Pitman, New York, NY.

Allegro, J. M. 1970. *The Sacred Mushroom and the Cross.* Hodder and Stoughton, London, U. K.

Ammirati, J. F., J. A. Traquair, and P. A. Horgen. 1985. *Poisonous Mushrooms of the Northern United States and Canada.* University of Minnesota Press, Minneapolis, MN.

Anon. 1989. *Amanita phalloides* poisoning. *Mycofile.* The Newsletter of the Vancouver Mycological Society: Issue MXXXIII, Jan. '89.

Baer, H. 1979. The Poisonous Anacardiaceae. In: Kinghorn, A. D. (ed.) *Toxic Plants.* Columbia University Press, New York, NY.

Bailey, L. H. Hortorium, Staff of. 1976. *Hortus Third. A Concise Dictionary of Plants Cultivated in the United States and Canada.* L. H. Bailey Hortorium, Cornell University. Macmillan Publishing Co., Inc., New York, NY.

Baginski, S., and J. Mowszowicz. 1963. *Krajowe Rosliny Trujace.* Panstwowe Wydawnictwo Naukowe, Lodz, Poland.

Bandoni, R. J. and A. F. Sczcawinski. 1976. *Guide to Common Mushrooms of British Columbia.* British Columbia Provincial Museum Handbook No. 24, Victoria, B.C.

Bergner, P. 1989. Comfrey, Coltsfoot, and pyrrolizidine alkaloids. *Medical Herbalism; a Clinical Newsletter for the Herbal Practitioner* 1 (1): 1, 3–5.

Brinker, F. J. 1983. *An Introduction to the Toxicology of Common Botanical Medicinal Substances.* National College of Naturopathic Medicine, Portland, OR.

Bruce, E. A. 1927. Astragalus campestris *and other Stock Poisoning Plants of British Columbia.* Canada Department of Agriculture, Bulletin No. 88, Ottawa, Ont.

Buck, R. W. 1961. Mushroom poisoning since 1924 in the United States. *Mycologia,* 53:538.

Canada Department of Agriculture. 1968. *Poison Ivy.* Publication No. 820, Ottawa, Ont.

Canadian Pharmaceutical Association. 1984. *Poison Management Manual.* (Rev.). Ottawa, Ont.

Ciegler, A. 1975. Mycotoxins: occurrence, chemistry, biological activity. *Lloydia* 38: 21–35.

Chi-Kit W., T. Johns, and G. H. N. Towers. 1980. Phototoxic and antibiotic activities of plants of the Asteraceae used in folk medicine. *Journal of Ethnopharmacology,* 2:279–290.

Claus, E. P., V. E. Tyler, and L. R. Brady. 1970. *Pharmacognosy.* Lea & Febiger, Philadelphia, PA.

Cooper, M. R. and A. W. Johnson. 1984. *Poisonous Plants in Britain and Their Effects on Animals and Man.* Ministry of Agriculture Fisheries and Food, Reference Book 161. Her Majesty's Stationery Office, London, U. K.

DeWolf, G. P. 1974. Guide to potentially dangerous plants. *Arnoldia* 34 (2): 45–91.

Der Marderosian, A. and F. C. Roia, Jr. 1979. Literature review and clinical management of household ornamental plants potentially toxic to humans. In: Kinghorn, A. D. (ed.) *Toxic Plants.* Columbia University Press, New York, NY.

Duke, J. A. 1985. *CRC Handbook of Medicinal Herbs.* CRC Press, Inc., Boca Raton, FL.

Emboden, W. 1979 (rev.) *Narcotic Plants. Hallucinogens, Stimulants, Inebriants, and Hypnotics, their Origins and Uses.* Collier Books, New York, NY.

Faulstich, H. 1980. Mushroom poisoning. *Lancet* II: 794–795.

Faulstich, H., B. Kommerell, and T. Wieland (eds.) 1980. *Amanita Toxins and Poisoning.* Verland Gerhard Witzstrock, Baden-Baden, Germany.

Fernald, M. L., Kinsey, A. C., and R. C. Rollins. 1958. *Edible Wild Plants of Eastern North America.* Harper & Row, New York, NY.

Floersheim, G. L. 1987. Treatment of human amatoxin mushroom poisoning. Myths and advances in therapy. *Medical Toxicology* 2: 1–9.

Fowler, M. H. 1980. *Plant Poisoning in Small Companion Animals.* Ralson Purina Co., St. Louis, MS.

Fuller, T. C. and E. McClintock. 1986. *Poisonous Plants of California.* University of California Press, Berkeley, CA.

Groves, J. W. 1979. *Edible and Poisonous Mushrooms of Canada.* Agriculture Canada Publication No. 1112, Ottawa, Ont.

Haard, R. and K. Haard. 1980. *Poisonous and Hallucinogenic Mushrooms.* Homestead Book Company, Seattle, WA.

Hall, R. L. 1973. Toxicants occurring naturally in spices and flavors. In: National Academy of Sciences. *Toxicants Occurring Naturally in Foods.* Washington, DC.

Hamon, N. W., J. W. Blackburn, and C. A. De Gooijer. 1986. *Poisonous Plants in the Home and Garden.* (authors all College of Pharmacy, University of Saskatchewan.) Miratext Enterprises Ltd., Winnipeg, Man.

Hardin, J. W. and J. M. Arena. 1974 (2nd ed.). *Human Poisoning from Native and Cultivated Plants.* Duke University Press, Durham, NC.

Hatfield, G. M. 1979. Toxic Mushrooms. In: Kinghorn, A. D. (ed.) *Toxic Plants.* Columbia University Press, New York, NY.

Hatfield, G. M. and L. R. Brady. 1975. Toxins of higher fungi. *Lloydia* 38 (1): 36–55.

Hedrick, U. P. (ed.) 1972. *Sturtevant's Edible Plants of the World.* Dover Publications, Inc., New York, NY.

Hessayon, D. G. 1985. *The Indoor Plant Spotter.* pbi Publications, Britannica House, Herts, England.

Hill, R. J. 1986. *Poisonous Plants of Pennsylvania.* Pennsylvania Department of Agriculture, Harrisburg, PA.

Hotson, J. W. and E. Lewis. 1934. *Amanita pantherina* of Western Washington. *Mycologia,* 26:384.

Howard, R. A. 1974. *Poisonous Plants.* Arnoldia, Harvard University, Jamaica Plain, MA.

Hulbert, L. C. and F. W. Oehme. 1984. *Plants Poisonous to Livestock: Selected Plants of the United States and Canada of Importance to Veterinarians.* Kansas State University Printing Service, Manhattan, KS.

Johns, T. and I. Kubo. 1988. A survey of traditional methods employed for the detoxification of plant foods. *Journal of Ethnobiology* 8 (1): 81–129.

Keeler, R. F. 1979. Toxins and teratogens of the Solanaceae and Liliaceae. In: Kinghorn, A. D. (ed.) *Toxic Plants.* Columbia University Press, New York, NY.

Keeler, R. F., K. R. Van Kampen, and L. F. James (eds.) 1978. *Effects of Poisonous Plants on Livestock.* Symposium on Poisonous Plants, Utah State University. Academic Press, New York, NY.

Kinghorn, A. D. (ed.) 1979. *Toxic Plants.* Columbia University Press, New York, NY.

Kinghorn, A. D. 1979. Cocarcinogenic irritant Euphorbiaceae. In: Kinghorn, A. D. (ed.) *Toxic Plants.* Columbia University Press, New York, NY.

Kingsbury, J. M. 1964. *Poisonous Plants of the United States and Canada.* Prentice-Hall, Inc., Englewood Cliffs, NJ.

Kingsbury, J. M. 1967. *Deadly Harvest. A Guide to Common Poisonous Plants.* George Allen and Unwin Ltd., London, U. K.

Kingsbury, J. M. 1979. The problem of poisonous plants. In: Kinghorn, A. D. (ed.) *Toxic Plants.* Columbia University Press, New York, NY.

Kramer, J. 1977. *A Complete Guide to Indoor Trees.* Coles Publishing Co. Ltd., Toronto, Ont.

Krochmal, A., L. Wilkins, D. van Lear, and M. Chien. 1974. Mayapple (*Podophyllum peltatum* L.). United State Department of Agriculture Research Paper NE–296. Northeast Forest Experiment Station, Upper Darby, PA.

Kuhnlein, H. V. and N.J. Turner. 1987. Cow-parsnip (*Heracleum lanatum* Michx.): an indigenous vegetable of native people of northwestern North America. *Journal of Ethnobiology* 6 (2): 309–324.

Lampe, K. F. and R. Fagerström. 1968. *Plant Toxicity and Dermatitis: a Manual for Physicians.* Williams and Wilkins, Baltimore, MD.

Lampe, K. F. and M. A. McCann. 1985. *AMA Handbook of Poisonous and Injurious Plants.* American Medical Association, Chicago, IL.

Leeuwenberg, A. J. M. (compiler). 1987. *Medicinal and Poisonous Plants of the Tropics.* Pudoc Wageningen, The Netherlands.

Levy, C. K. and R. B. Primack. 1984. *A Field Guide to Poisonous Plants and Mushrooms of North America.* The Stephen Greene Press, Brattleboro, VT.

Lewis, W. H. 1979a. Poke root herbal tea poisoning. *Journal of the American Medical Association* 242: 2759.

Lewis, W. H. 1979b. Snowberry (*Symphoricarpos*) poisoning in children. *Journal of the American Medical Association* 242: 2663.

Lewis, W. H. and M. P. F. Elvin-Lewis. 1977. *Medical Botany. Plants Affecting Man's Health.* John Wiley & Sons, New York, NY.

Lewis, W. H., P. Vinay, and V. E. Zenger. 1983. *Airborne and Allergenic Pollen of North America.* The Johns Hopkins University Press, Baltimore, MD.

Lincoff, G. H. 1981. *The Audubon Society Field Guide to North American Mushrooms.* Alfred A. Knopf, New York, NY.

Lincoff, G. H. and D. H. Mitchel. 1977. *Toxic and Hallucinogenic Mushroom Poisoning. A Handbook for Physicians and Mushroom Hunters.* Van Nostrand Reinhold Co., New York, NY.

McKenny, M. 1971. *The Savory Wild Mushroom* (rev. and enlarged by D. E. Stuntz). University of Washington Press, Seattle, WA.

McLean, A. and H. H. Nicholson. 1958. *Stock Poisoning Plants of the British Columbia Ranges.* Canada Department of Agriculture, Publ. No. 1037, Ottawa, Ont.

McPherson, A. 1979. Pokeweed and other lymphocyte mitogens. In: Kinghorn, A. D. (ed.) *Toxic Plants.* Columbia University Press, New York, NY.

Madlener, J. C. 1977. *The Seavegetable Book.* Clarkson N. Potter, Inc., New York, NY.

Meijer, W. 1974. *Podophyllum peltatum,* May Apple. A potential new cash-crop plant of eastern North America. *Economic Botany* 28: 68–72.

Miller, J. A. 1973. Naturally occurring substances that can induce tumors. In: National Academy of Sciences. *Toxicants Occurring Naturally in Foods.* Washington, DC.

Miller, O. K. 1980. *Mushrooms of North America.* E. P. Dutton & Co., Inc., New York, NY.

Millspaugh, C. F. 1974. *American Medicinal Plants* (originally published in 1892 under the title *Medicinal Plants*). Reprinted by Dover Publications, Inc., New York, NY.

Mitchell, J. and A. Rook. 1979. *Botanical Dermatology. Plants and Plant Products Injurious to The Skin.* Greengrass Ltd., Vancouver, B.C.

Morton, J. F. 1958. Ornamental plants with poisonous properties. *Proceedings of the Florida State Horticultural Society* 71: 372–380; 75: 484–491 (1962).

Morton, J. F. 1971. *Plants Poisonous to People.* Hurricane House, Miami, FL.

National Academy of Sciences. 1973. *Toxicants Occurring Naturally in Foods.* Washington, DC.

National Safety Council. 1975. Poison ivy, poison oak, and poison sumac. Data Sheet 304, Revision A. *National Safety News,* September, 1975:99–102.

Patterson, D. S. P. 1982. Mycotoxins. *Environmental Chemistry* 2: 205–233.

Patwardhan, V. N. and J. W. White. 1973. Problems associated with particular foods (includes discussion on honey toxins). In: National Academy of Sciences. *Toxicants Occurring Naturally in Foods.* Washington, DC.

Richardson, D. H. S. 1975. *The Vanishing Lichens.* David & Charles, London, U. K.

Rumack, B. H. and E. Salzman (eds.) 1978. *Mushroom Poisoning: Diagnosis and Treatment.* CRC Press, Inc., Boca Raton, FL.

Schantz, E. J. 1973. Seafood Toxicants. In: National Academy of Sciences. *Toxicants Occurring Naturally in Foods.* Washington, DC.

Scheel, L. D. 1973. Photosensitizing agents. In: National Academy of Sciences. *Toxicants Occurring Naturally in Foods.* Washington, DC.

Schneider, D. 1986. Poisonous plants. *Canadian Geographic* 106 (2): 60–65.

Schultes, R. E. and A. Hofmann. 1980. *The Botany and Chemistry of Hallucinogens.* Charles C. Thomas, Springfield, IL.

Smith, A. K. 1921. *Lichens.* Cambridge University Press, London, U. K.

Soper, J. H. and M. L. Heimburger. 1982. *Shrubs of Ontario.* Royal Ontario Museum, Toronto, Ont.

Stamets, P. E. 1978. *Psilocybe Mushrooms & their Allies.* Homestead Book Company, Seattle, WA.

Sweet, M. 1962. *Common Edible and Useful Plants of the West.* Naturegraph Company, Healdsburg, CA.

Szczawinski, A. F. and N.J. Turner. 1980. *Wild Green Vegetables of Canada.* National Museum of Natural Sciences, National Museums of Canada, Ottawa, Ont.

Towers, G. N. 1979. Contact hypersensitivity and photodermatitis evoked by Compositae. In: Kinghorn, A. D. (ed.) *Toxic Plants.* Columbia University Press, New York, NY.

Tu, A. T. (ed.) 1988. *Handbook of Natural Toxins.* Vol. 3. *Marine Toxins and Venoms.* Marcel Dekker Inc., New York, NY.

Turner, N.J. 1975. *Food Plants of British Columbia Indians.* Part 1. *Coastal Peoples.* British Columbia Provincial Museum Handbook No. 34, Victoria.

Turner, N.J. 1977. Economic importance of black tree lichen (*Bryoria fremontii*) to the Indians of western North America. *Economic Botany* 31: 461–470.

Turner, N.J. 1978. *Food Plants of British Columbia Indians.* Part 2. *Interior Peoples.* British Columbia Provincial Museum Handbook No. 36, Victoria.

Turner, N.J. 1982. Traditional use of devil's-club (*Oplopanax horridus;*

Araliaceae) by Native peoples in western North America. *Journal of Ethnobiology* 2 (1): 17–38.

Turner, N.J. 1984. Counter-irritant and other medicinal uses of plants in Ranunculaceae by Native peoples in British Columbia and neighbouring areas. *Journal of Ethnopharmacology* 11: 181–201.

Turner, N.J., R. Bouchard, and D. I. D. Kennedy. 1980. *Ethnobotany of the Okanagan-Colville Indians of British Columbia and Washington.* British Columbia Provincial Museum Occasional Paper No. 21, Victoria.

Tyler, V. E. 1987 (rev.) *The Honest Herbal. A Sensible Guide to Herbs and Related Remedies.* George F. Stickley Co., Philadelphia, PA.

Wasson, R. 1968. *"Soma": Divine Mushroom of Immortality.* Harcourt, Brace and World. New York, NY.

Wyllie, T. D. and L. G. Morehouse (eds.) 1978. *Mycotoxic Fungi, Mycotoxins and Mycotoxicoses. An Encyclopedic Handbook,* Vol. 2. Marcel Dekker Inc., New York, NY.

INDEX

Note: Italicized numbers refer to illustrations.